Introduction to Integrated Pest Management

Introduction to Integrated Pest Management

Mary Louise Flint

University of California
Davis, California

and

Robert van den Bosch

University of California
Berkeley, California

PLENUM PRESS • NEW YORK AND LONDON

Library of Congress Cataloging in Publication Data

Flint, Mary Louise, 1949-
Introduction to integrated pest management.

Bibliography: p.
Includes index.
1. Pest control, Integrated. I. Van den Bosch, Robert, joint author. II. Title.
SB950.F56 628.9'6'068 80-28479
ISBN 0-306-40682-9

A Division of Plenum Publishing Corporation
233 Spring Street, New York, N.Y. 10013

Printed in the United States of America

Foreword

Integrated control of pests was practiced early in this century, well before anyone thought to call it "integrated control" or, still later, "integrated pest management" (IPM), which is the subject of this book by Mary Louise Flint and the late Robert van den Bosch. USDA entomologists W. D. Hunter and B. R. Coad recommended the same principles in 1923, for example, for the control of boll weevil on cotton in the United States. In that program, selected pest-tolerant varieties of cotton and residue destruction were the primary means of control, with insecticides considered supplementary and to be used only when a measured incidence of weevil damage occurred. Likewise, plant pathologists had also developed disease management programs incorporating varietal selection and cultural procedures, along with minimal use of the early fungicides, such as Bordeaux mixture. These and other methods were practiced well before modern chemical control technology had developed.

Use of chemical pesticides expanded greatly in this century, at first slowly and then, following the launching of DDT as a broadly successful insecticide, with rapidly increasing momentum. In 1979, the President's Council on Environmental Quality reported that production of synthetic organic pesticides had increased from less than half a million pounds in 1951 to about 1.4 billion pounds—or about 3000 times as much—in 1977. At the same time ecologically oriented methods of pest control were largely replaced by reliance on pesticides. Although serious weaknesses in a policy of sole reliance on chemicals had surfaced, the benefits in lives saved, improved human health, and increases in food and fiber production from use of the new insecticides were strong arguments against a return to a broader, more ecologically based approach.

In 1959, the concept of IPM received a major thrust in the formal definition, statement of principles, and description of the necessary techniques proposed by a group of California workers—V. M. Stern, Ray F. Smith, Robert van den Bosch, and K. S. Hagen. In that paper, the consequences of a policy of sole reliance on insecticides were, in effect, predicted and the weaknesses of such a strategy for crop protection illustrated. The authors noted that the destruction of natural enemies by insecticides has a twofold effect on future pest problems. First, the natural suppressive effects of the enemies of the target pest species are reduced, allowing rapid target pest resurgence. Second, the reduced action of natural enemies on other innocuous or minor pests can unleash these to become major pests. An even more significant factor forcing consideration of integrated control was the widespread development of resistance to the materials, one after another, in the target pests themselves. Entomologists began to see that soon they might be unable to control many insects by the use of chemicals.

Concomitant with these purely economic entomological concerns was a growing awareness of the dangers to man, indeed to the whole biosphere, from extensive unrestrained use of pesticides. Rachel Carson's *Silent Spring*, published in 1962, not only documented some of these crop-centered problems with chemical pest control, but also dared to consider the broader implications of their unrestrained use for human health and the well-being of all organisms on earth. Events moved rapidly after *Silent Spring* served to crystallize in the public mind and in the halls of Congress a consciousness of the dangers inherent in an unrestrained release of deadly chemicals into the environment. The list of chemicals soon included not only pesticides but also widely distributed air, water, and soil pollutants in great variety and, in addition, chemical and radioactive wastes that had been "stored away" in various concentrated caches and largely forgotten.

In the late 1960s and the early 1970s, both the USDA and the National Science Foundation (the latter joined later by the Environmental Protection Agency) launched efforts to look at insect pest control in a more holistic way—that is, to concentrate on a given crop or cropping system in a limited farming area—and to seek a thorough understanding of the system, especially in regard to crop growth, pest impact, and cost–benefit relationships. This, in essence, is simply putting into practice what IPM specialists had been saying since 1959 (Stern *et al.*), and even as early as 1923 (Hunter and Coad). These efforts have now been underway on a large scale for a decade and have more recently been materially broadened to consider for a few major crops, in a more balanced way, the whole complex of crop pests and agronomic considerations.

The great expansion of effort in this latter activity made it apparent that different people talking about IPM often have entirely different ideas about what it is, how it can be done, and what can be expected of it. Moreover, it became evident that if IPM were to become the "norm" for pest control in this country (or elsewhere) a large force of field people knowledgeable of its principles and working concepts and techniques would be required. An educational program of extensive influence would be needed.

Flint and van den Bosch's *Introduction to Integrated Pest Management* fills a real void in this educational sphere. It is written in clear, succinct language easily followed by the beginner—the student and trainee in IPM. Yet it competently and persuasively deals with the subtle and complex interacting relationships bearing upon agriproduction, pest control, costs and benefits, and external consequences concerning public health and environmental quality.

The book starts with an explanation of the place of man, pests, and other organisms and their evolution in the scheme of things (Chapter 1). In Chapter 2, it illustrates how man's release of pesticides can have unintended adverse effects in the fields themselves and in areas far removed from these fields (or forests, marshes, or livestock sheds). These first chapters furnish the minimal background information necessary for the student who wishes to understand the IPM strategy and why it is needed.

The book then deals with the basic IPM concepts themselves (Chapter 3)—with the concept of "pest" and with the natural factors operating to keep potential pests in check—including excellent illustrations of how a pesticide may disturb such natural checks.

There is an extended and intriguing discussion (Chapter 4) of the earliest of man's efforts to control pests. These efforts go back a few thousand years; this is so even for chemicals (sulfur). The sketches of early spray equipment will seem archaic to the modern spray user. This account extends right up to the present, with detailed explanations and sketches illustrating the problems arising with the use of modern insecticides and with new developments in IPM itself.

Chapter 5 focuses on the economic, social, and environmental costs of pest control and on the various conflicting interests and values emanating from man's attempts to manage agroecosystems for his own benefit. The chapter deals with inputs of fertilizer, water, pesticides, and labor and with pest upsets; with the development of pesticide-resistant pests; with the persistence of pesticides in the soil, water and air and their accumulations in organisms on higher levels of the food chain; and with direct public health problems.

Chapters 6 and 7 discuss in detail the concept of IPM itself, how to establish a program, the tools for decision-making, and the role of the IPM manager. Chapter 8 considers a number of cases where IPM has been employed. These include alfalfa in California, apples in Nova Scotia, flue-cured tobacco in North Carolina, cotton in California's San Joaquin Valley, and mosquitoes in a salt marsh in California. Chapter 9 points out the important role of the IPM consultant or specialist in the implementation of such programs.

The final chapter deals with many of the indefinite and complicated technical, political, and economic circumstances that would seem to determine whether or not IPM, as defined here, will emerge as the norm for pest control in the United States.

Very recently, as a consequence of preliminary results from research and an extensive series of pilot tests of IPM systems for several crops in various sections of the country, IPM has "caught on" as the thing to do. The President, Congress, and at least one state legislature (in California) have urged adoption of IPM in one way or another. In his 1979 Environmental Message, President Carter instructed the Council on Environmental Quality "to recommend actions which the federal government can take to encourage the development and application of techniques to be used for sound IPM programs." The State of California has decreed by law that IPM must be employed "wherever feasible." To be consistent with the President's wishes, those endeavoring to use IPM should view it as defined by the President in his 1979 Environmental Message:

> IPM uses a system approach to reduce pest damage to tolerable levels through a variety of techniques, including natural predators and parasites, genetically resistant hosts, environmental modifications and, when necessary and appropriate, chemical pesticides. IPM strategies generally rely first upon biological defenses against pests before chemically altering the environment.

The President further directed the agencies of government "to modify . . . programs to support and adopt IPM strategies wherever practicable."

In both the President's directive and that of California's legislature, the matter rests on what is feasible or practicable. This cannot be determined for many crops and situations without much more detailed research, field trials, and education at all levels. Flint and van den Bosch's book is a most important step in establishing the conceptual and educational background required for proper research, for the development of sound, practicable IPM programs, and for training personnel to be engaged in this effort.

Carl B. Huffaker
Division of Biological Control
Department of Entomological Sciences
University of California
Berkeley, California

Preface

The science and art of pest control has changed significantly in the last decade. The cost of pesticides has skyrocketed, the short-term and long-term effects of pesticides have become of general concern, and controlling pests has become much more difficult and complex. All these factors have sparked an increasing interest in integrated pest management (IPM) as a method of reducing the cost, increasing the efficacy, and lessening the hazards of pest control.

Despite this intense interest and the development of a few highly refined IPM systems, there has been a clear lack of information at the introductory level for the undergraduate student, nonspecialist teacher, grower, fieldman, pest control operator, consultant, or ordinary citizen who seeks more than the short articles on IPM in the popular or agricultural press. Much of the literature on specialized aspects of IPM has been available to the scientific community for a number of years, but for the most part it is unsuitable for readers who do not have an extensive background in ecology, entomology, pest control technology, and, sometimes, mathematics.

This book was developed to fill the need for a textbook that presents a comprehensive review of the basic principles and methods of IPM, presuming little or no previous background knowledge of the user in ecology, mathematics, economics, or pesticide chemistry. The purpose has been to give just enough background for the reader to gain insights into the contributions of these disciplines to IPM, and to then get on with descriptions and case histories illustrating the potential of IPM. IPM has been aptly called "applied ecology": pest control activities are planned and carried out in a way that makes maximum use of naturally occurring pest control factors and that is least disruptive to the overall managed ecosystem, yet is ever mindful of the economic realities under which growers, foresters, and pest control specialists must make decisions. The emphasis in this book is on the presentation of the ecological and economic considerations that constitute the bottom line of IPM. Ecological considerations are vital if unnecessary disruptions of natural controls, hazard to the environment, and rapid development of pesticide resistance are to be avoided. It is self-evident that economic considerations are also vital: the producer must make a profit.

No IPM program can function without the constant attention and vigilance of informed decision makers, variously called, for example, integrated pest control specialists, pest control advisers, scouts, or fieldmen. It has been our hope that this text will provide the necessary background for incorporating IPM concepts into field monitoring programs and pest control decision-making in the field. Of course, consultation with local university and extension experts is necessary to develop the details of a specific IPM program suitable for practical application to a given crop in any given area. It has been our hope also that this book will find a place in the general education of our citizenry so that pesticide usage and pest control methods can be seen in a broader perspective than is conveyed by the popular press or the profit-motive advertising of pesticide companies.

This book has been written by two entomologists and it may be criticized for its entomological emphasis while carrying a general title implying all classes of pests. This weakness is due, unfortunately, as much to the limited development of IPM programs that encompass weed, vertebrate, and pathogen problems as to the limited (that is, entomological) experience of the authors. We are confident that within the next decade the principles described in this book will be found to be more widely applicable to the management of other classes of pests and that future books on this subject will have more examples and material to draw upon from these other pest control disciplines. The specific guidelines and techniques needed for integrated pest management of weeds, plant pathogens, nematodes, and vertebrates, on the other hand, will continue to be quite different, in many cases, from those used in the management of insects and mites.

This book was first developed under a grant from the Office of Education, U.S. Department of Health, Education, and Welfare (Grant No. G007500907). Additionally, the authors are deeply indebted to a number of people who played key roles in the development of the book. First and foremost is Carl Huffaker, whose advice, thorough review, and support were essential in preparing the final manuscript. A. P. Gutierrez, Bill and Helga Olkowski, Christine Merritt, and a number of other people at the Division of Biological Control of the University of California, Berkeley, contributed ideas and support. Joanne Fox typed the original manuscript and Nettie Mackey and Jane Clarkin spent countless hours over the last few years related to the production of the book. The illustrations in this book are the work of Lisa Haderlie, Randy Elliot, and Naomi Schiff. Their ideas, good humor, and willingness to meet deadlines were a blessing to the authors.

Last, I must note here the untimely death of my collaborating author and inspiration, Professor Robert van den Bosch, who died prior to the completion of the final manuscript.

Mary Louise Flint

Davis, 1981

Contents

CHAPTER 1

Man, Pests, and the Evolution of IPM

An Introduction

The definition of "pest" is totally human-oriented. Organisms designated *pests* compete with people for food, fiber, and shelter; transmit pathogens; feed on people; or otherwise threaten human health, comfort, or welfare. It could be said that, previous to the appearance of humans, there were no pests—just millions of different organisms struggling for survival; the arrival of humans and the continuing development of the human life-style have provided the sole basis for labeling an ever-increasing number of these surviving organisms "pests."

A strictly ecological viewpoint would consider every link in the food chain (or each of these differently adapted organisms) as equally important in an ecosystem's assemblage of plants, animals, and interacting physical environment. To such pure ecologists, describing an organism as a negative factor (or a "pest") just because its link in the food chain happens to be at the same spot as or right next to our own would be unthinkable!

But in human-oriented sciences—e.g., economics, medicine, agriculture, silviculture, and park and recreation area management—which aim to insure human survival and enhance our life-style, the role of such competing or consuming organisms, whose appetites or habits might limit our "success," can be considered nothing short of adverse. When, why, how, and to what degree various organisms become pests, however, is a matter of considerable debate in all these sciences.

Organisms that have become "pests" are not limited to any class, phylum, or even kingdom. They are as varied as the habits that make them undesirable. Insects are frequent pests (and it is no wonder, since they make up more than 75% of the world's animal species!) (Figure 1-1). A number of mite, tick, nematode, mollusc, and other invertebrate species have become pests. Vertebrates, including rodents, deer, coyotes, and birds, may become serious pests in some situations. Microorganisms (e.g., bacteria, fungi, protozoa, rickettsiae, viruses, and mycoplasmas), particularly those that are pathogenic to important plants and animals, cause many problems. Weeds—ordinary plants in places where they are not wanted—comprise another category of common pestiferous organisms.

In the human war against pests, our battle strategies and tactics have evolved through the ages, becoming more sophisticated and, for the most part, more effective. Our first methods of pest control were undoubtedly the hand-picking, swatting, and squashing of insects and other small invertebrates. Later we learned how to manipulate the environment so that it became less favorable to pests; examples of these early environmental modifications include flooding or burning fields to destroy weed, insect, and other invertebrate pests, and using scarecrows to keep birds away. The utilization of natural enemies to control pest organisms dates back several thousand years; but it was the rather recent application of such "biological control" methods against the devastating cottony cushion scale insect that truly demonstrated the value of this approach to pest control. In this case, an imported lady beetle saved the California citrus industry from total ruin in the late nineteenth century.

Materials with pesticidal properties, such as plant-derived chemicals (e.g., pyrethrum) and arsenic and sulfur, were used sporadically and largely ineffectively from the time of the Greek and Roman Empires. More sophisticated use of pesticides evolved in the latter half of the nineteenth century after two copper-based fungicides, Bordeaux mixture and Paris Green, were found to be effective against mildews and other diseases that were threatening the grape industry in France. Both fungicides, especially Paris Green, which contained arsenic in addition to copper, turned out to be useful as insecticides as well and were subsequently used regularly to kill a spectrum of insect pests.

The years preceding World War II saw increasing use of these and other chemical control materials against arthropod pests and plant diseases. However, because of the hazardous nature of these pesticides, their expense, the poor application techniques available, and the ineffectiveness of the materials in many situations, chemical pest control remained a limited technology, and pest management still depended largely on environmental manipulation, sanitary practices, biological control, and, to a certain degree, luck.

But then, the discovery of the insecticidal properties of DDT in the 1940s changed all that. Here was an apparent "miracle" pesticide—cheap, incredibly effective at low dosages, long-lasting, easy to apply, and lethal to an unprecedented spectrum of insect pests. Soon other chlor-

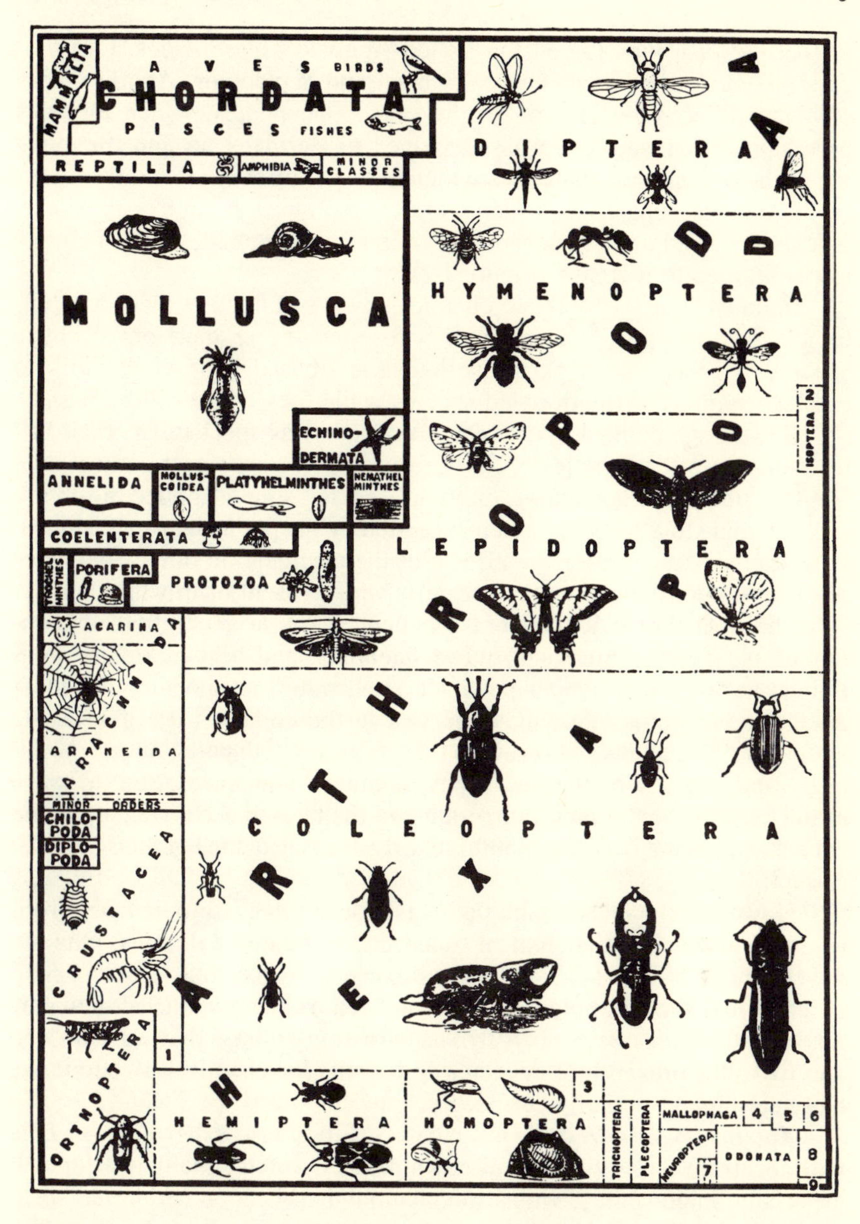

FIGURE 1-1. A diagram to show the relative size in number of known species of various animal phyla. The overwhelming prevalence of the Arthropoda in the animal kingdom, and particularly the insects (class Hexapoda here), is clear (note the small size of the mammals, for instance).

In the diagram, number 1 represents minor classes of the Arthropods. Other numbers represent small orders of insects: 2, the fleas; 3, the book-lice, bark-lice, and dust-lice; 4, the mayflies; 5, the earwigs; 6, the thrips; 7, the bloodsucking lice, stylops, and scorpion flies; 8, the springtails and snow-fleas; 9, the silverfish (from Metcalf and Flint, 1951).

inated hydrocarbons joined the list of "miracle" insecticides, e.g., lindane, dieldrin, methoxychlor, chlordane, and heptachlor. Another class of highly effective insecticides, the organophosphates, which includes parathion and malathion, was developed in Germany around the same time. The organophosphates were then followed by the carbamates. Modern herbicides, fungicides, rodenticides, and other pest control chemicals quickly followed on the heels of the "miracle" insecticides, and their use has continued to increase (Figure 1-2).

Initially, control by these chemicals was so effective that some entomologists and pest managers even foresaw the eradication of entire species of pests and in partial seriousness advised young collectors to catch specimens of the doomed species while they could still be found.

But though control by the new pesticides was spectacular, their utilization was often mindless. The grower, government pest-control specialist, commercial applicator, or home gardener simply applied the chemical according to a schedule (often suggested by the pesticide manufacturer). Few stopped to consider the effects of the pesticide on other organisms or even to determine whether the pests were present or in what density.

The post-World War II era produced a whole generation of entomologists, plant pathologists, weed scientists, and pest managers well-trained in the art of proper pesticide application, in choosing the most potent and/or economical material, and in the complexities of spraying equipment. Frequently, however, they were less educated about the biology and ecology of the target pest organism and were often unaware of the natural control factors operating in the treated ecosystem or of the nature and identity of other nontarget organism casualties of their biocidal assault.

However, the DDT "miracle" was indeed too good to be true. Problems did arise. The first hint of modern insecticides' fallibility was the development of resistance in insects exposed to the toxic materials. In other words, some populations that had been frequently doused with particular insecticides developed strains (*resistant strains*) that were able to survive in the presence of even heavy doses of the chemicals— becoming more serious pests than ever before.

The post-World War II pesticide revolution also ushered in a whole new spectrum of previously unknown pests. Suddenly many arthropods, especially spider mites, whose populations had been generally small or moderate, became major pests of crops and other resources. For the most part this sudden leap to prominence was prompted by the insecticides' destruction of natural enemies that had previously held potentially injurious species under restraint. Freed of their natural enemies and tolerant of the pesticides, these organisms survived and multiplied with incredible speed.

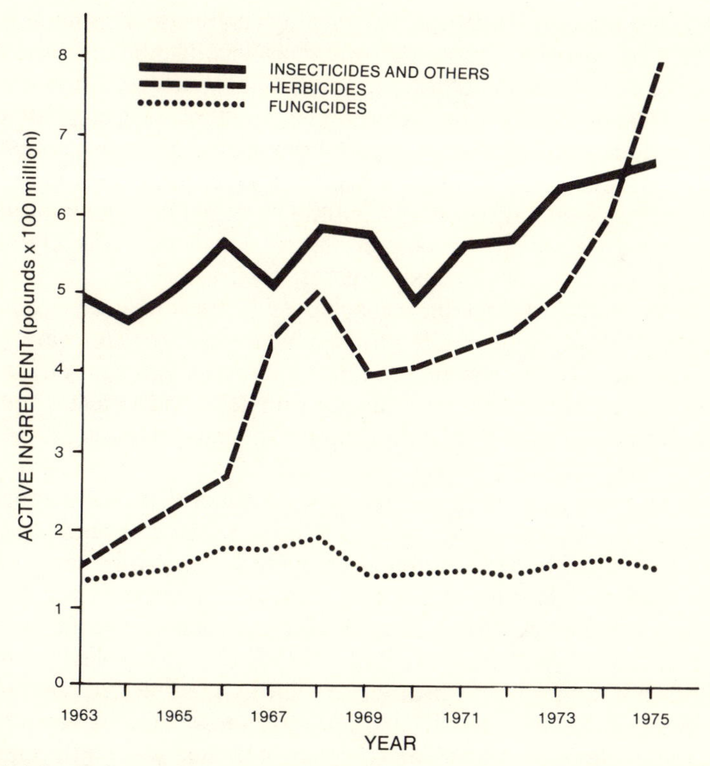

FIGURE 1-2. U.S. production of synthetic organic insecticides, herbicides, and fungicides, 1963–1975.

Environmental contamination and the killing of wildlife were perhaps the greatest tragedies of the widespread overuse of pesticides. Honeybees and fish, birds, and other wild animals were the innocent and often unknown victims of the massive pesticides spraying of agricultural crops and forest and recreation areas. Unlike pest insects, many of these organisms had neither the reproductive capacity nor the short generation time required to develop resistance. Reports of death and injury among farmworkers and other human victims imprudently exposed to toxic dosages of pesticides were numerous and sparked the enactment of laws regulating and even prohibiting the use of many of these toxic substances.

Yet many of the negative side effects resulting from the overdependence on pesticides should not have been surprising to twentieth-century scientists and could have been predicted had thoughtful biologists, ecologists, public health specialists, and geneticists been closely involved and on the lookout for possible problems. For instance, the development of insecticide resistance among important agricultural pests had been documented as early as 1912, and warnings about the serious problems that could result from such insecticide resistance had been sounded well before the advent of the synthetic organic insecticides. Such a speedy evolution of tolerance to new environmental stresses is a major reason that many rapidly reproducing invertebrates have been so successful; it is a process that follows well-known laws of natural selection. As for environmental contamination and the killing of wildlife, pollinators, and important natural enemies, these too could have been predicted by biologists, ecologists, toxicologists, and others familiar with the basic concepts of food chains and the biosphere's biogeochemical cycles. Yet somehow these questions (in fact raised by a few scientists) were pushed aside, and most pest managers and pest control researchers of the 1940s and 1950s, mesmerized by the seeming simplicity and efficiency of complete reliance on pesticidal control, forgot the laws of ecology and stumbled into chaos.

Thus, it is of critical importance to remember that pest management is basically an ecological matter. Man wants to secure as much of a given resource as possible with minimum competition from other organisms in the ecosystem. Therefore, effective pest management must begin with an ecological outlook. Artificial controls (e.g., pesticides and cultural practices) must be looked on as tools to be fitted into the environment as unobtrusively and with as little disruption as possible. It is imperative that we consider the effects of the various control actions both on each other and on the rest of the environment, and thus prevent the negation of one pest control factor by another (e.g., biological control by chemical control).

In recent years a new comprehensive approach to pest control has been developing; it is termed *integrated pest management* or *integrated control*. Integrated pest management (IPM) is an ecologically based pest control strategy that relies heavily on natural mortality factors such as natural enemies and weather and seeks out control tactics that disrupt these factors as little as possible. IPM uses pesticides, but only after systematic monitoring of pest populations and natural control factors indicates a need. Ideally, an integrated pest management program considers all available pest control actions, including no action, and evaluates the potential interaction among various control tactics, cultural practices, weather, other pests, and the crop to be protected. Under IPM, natural enemies, cultural practices, resistant crop and livestock varieties, microbial agents, genetic manipulation, messenger chemicals (such as sex attractants), and pesticides become mutually augmentative instead of individually exclusive or even antagonistic—as has been so often the case under pesticide-dominated control.

An integrated pest management program is comprised of six basic elements: (1) people: the system devisers and pest managers; (2) the knowledge and information necessary to devise the system and make sound management decisions; (3) a program for monitoring the numbers and state of the ecosystem elements—e.g., resource, pest, and natural enemies; (4) decision-making levels: the pest densities at which control methods are put into action; (5) IPM methods: the techniques used to manipulate pest populations; and (6) agents and materials: the tools of manipulation.

Integrated pest management systems are dynamic, as are the ecosystems in which they are invoked, and usually involve continuous information gathering and evaluation as the resource and its associated physical and biological environment go through their seasonal progressions. Thus, in insect IPM, control action programs evolve as and if pest problems develop; the preprogrammed rigidity present in conventional pesticide systems, where pest managers spray automatically according to a predetermined schedule, has been eliminated.

The integrated pest manager or adviser is the key to this dynamism. He/she must constantly "read" the situation, evaluate pest populations according to previous experience and knowledge, and initiate actions and choose appropriate materials and agents as conditions dictate. It is imperative that pest managers be aware of the ecological and biological effects of their actions at all times, not only on the target pest organisms, but also on natural enemies and nonpest organisms.

While the need for integrated pest management is now accepted by workers in most areas of pest control, effective and working programs have been implemented for only a dozen or so resources. These are mostly agricultural crops and are largely for the control of insect and mite pests. This text must consequently rely heavily on examples from agricultural crop insect pest situations. However, programs involving a spectrum of pest problems for livestock and in urban, forest, and recreation areas are currently being worked out; examples of these will be introduced in the text whenever possible.

CHAPTER 2

Human-Managed Environments as Systems within the Biosphere

A sound ecological understanding of pest species, of the managed environment, and of the environmental effects of pest management procedures is a prerequisite to the instigation of a successful integrated pest management program. Pest problems do not arise in a vacuum; they arise because a combination of factors in the environment favors the growth of pest populations. For instance, availability of food and water, favorable weather, shelter, and a shortage of the organisms that normally feed on the pest species (predators) may provide conditions conducive to a pest outbreak. Likewise, actions taken to control pests have effects on the surrounding environment beyond killing the target pest. This chapter briefly describes the major ecological principles most necessary to gain an insight into the causes of pest problems, to understand integrated pest management literature, and to manage pests in an ecologically sound manner. For a more detailed discussion of these principles, the reader is urged to consult the many fine ecology textbooks available.

Pest problems are biological phenomena. Pests are organisms occupying space, eating food, and/or carrying out other biological functions in places where, for some reason, we don't want them. As biological entities, pests can be and, in fact, must be viewed on several levels: (1) as genetically unique individuals struggling for survival; (2) as interbreeding populations of the same species occupying the same habitat; (3) as integral parts of communities of different kinds of organisms living in

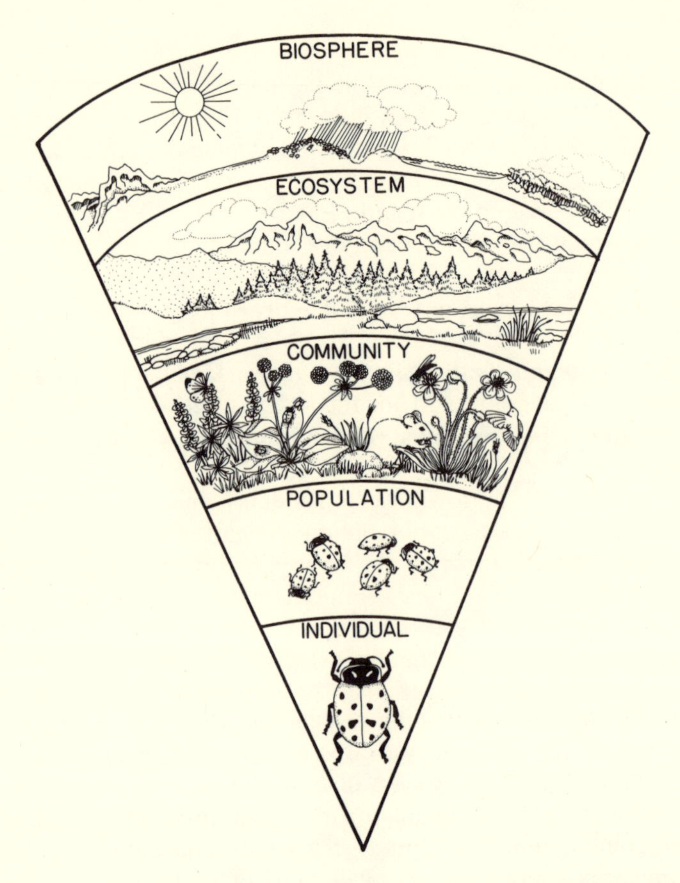

FIGURE 2-1. Levels of viewing biological activity. The individual in this case, the convergent lady beetle, is not a pest but an important predator of several pest insects.

certain areas, eating each other, and competing for food and shelter; (4) as parts of local ecosystems of organisms and their interacting physical environments; and (5) as parts of the biosphere—the total mass of living organisms on the earth's surface and its supporting nonliving environment (Figure 2-1).

THE BIOSPHERE

The *biosphere* is the largest biological unit, and for many life support systems, such as water transportation and distribution, the only unit in which the full process can be viewed. The biosphere concept recognizes that, because of the global nature of many of these interactions, physical

TABLE 2-1
A Partial Listing of the Essential Elements Required by the Biosphere's Organisms

Element	Example of function in living organisms
1. Hydrogen 2. Carbon	These two elements are present in every organic molecule
3. Oxygen	Essential for aerobic respiration
4. Nitrogen	Present in all proteins; the structural material of life
5. Phosphorus	Important in bone tissue and nerve and muscle regulation, and in transfer of energy within living systems
6. Sulfur	Important component of certain essential proteins
7. Potassium	Functions in transmission of nerve impulses and in protein biosynthesis
8. Magnesium	Important component of chlorophyll and bone; important in transmission of nerve impulses and in utilization of calcium and vitamin C
9. Sodium	Important in regulating osmotic pressure, which maintains cell shape and size; important in transmission of nerve impulses
10. Calcium	Major component of bones and teeth (as calcium phosphate); important in blood coagulation and muscle contraction
11. Iron	Functions in vertebrate blood (as hemoglobin) for oxygen transport and (as myoglobin) for storage; important in photosynthesis (as ferredoxin)
12. Copper	Functions in invertebrate blood (as hemocyanin) for oxygen transport; important in photosynthesis (as plastocyanin)
13. Zinc	Essential component of some enzymes, such as carbonic anhydrase (for CO_2 formation and regulation of acidity) and carboxypeptidase (for protein digestion); essential for synthesis of insulin
14. Manganese	Important in urea formation, lactation
15. Cobalt	Important in DNA biosynthesis; constituent of vitamin B_{12}
16. Molybdenum	Essential component of several important enzymes, including nitrogenase, the key enzyme of nitrogen-fixing soil bacteria
17. Chlorine	A key component of gastric juice and serum; participates in transport of CO_2 in blood and in regulating osmotic pressure, which maintains cell size and shape
18. Iodine	Important in function of thyroid gland in man, which regulates growth, development, and metabolic activities
19. Fluorine	Important in maintenance of teeth
20. Chromium	Function not yet clarified
21. Selenium	Function not yet clarified

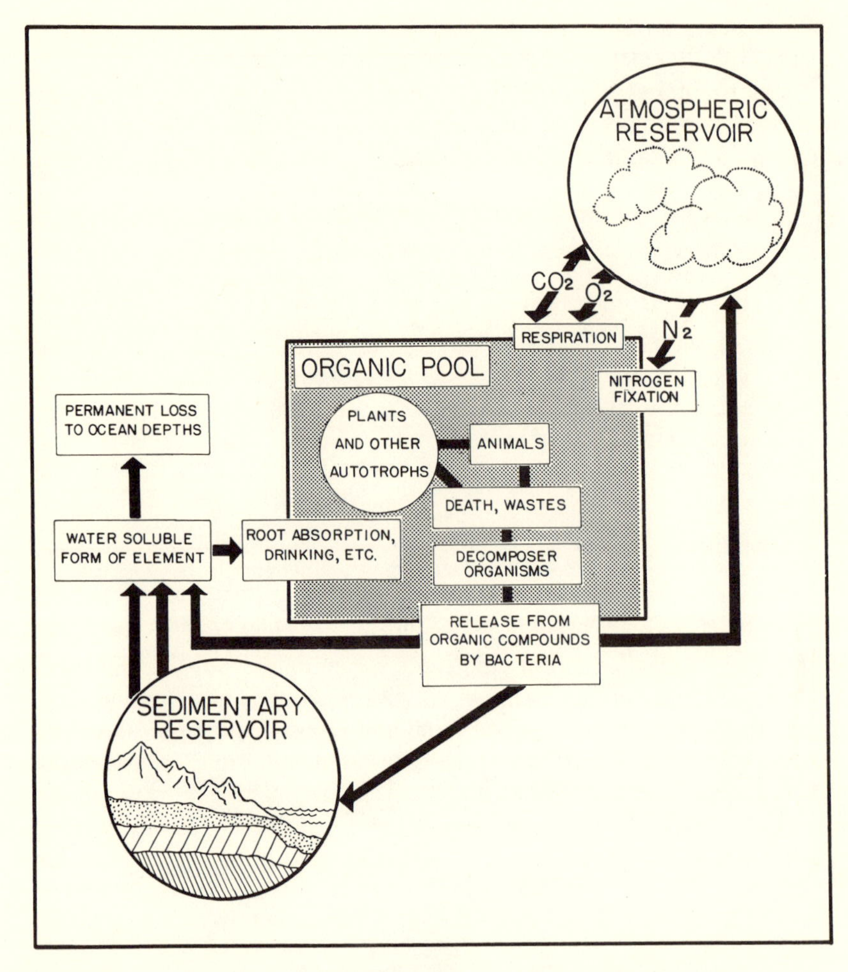

FIGURE 2-2. Generalized biogeochemical cycles: the pathways along which materials are acquired and disposed of by the biosphere's living and nonliving components.

or biological modification of the environment in one part of the world may affect organisms in very remote places.

Two forces unify the biosphere: (1) the cycling, transformation, and transport through the biosphere's living and nonliving components of the materials (e.g., hydrogen, oxygen, carbon, nitrogen, phosphorus, sulfur, potassium, magnesium, sodium, calcium, and the trace elements) that make up organisms' bodies; and (2) the process of capturing and distributing the energy (ultimately from the sun) that powers the activities of the biosphere's organisms. Without proper functioning of these two processes, life on earth would soon cease.

Approximately 30 essential elements are required for normal growth and body maintenance in most organisms. The most important of these elements and examples of their functions are listed in Table 2-1. Organisms acquire these materials via specified pathways: (1) from nonliving sedimentary reservoirs on the earth's surface, usually in a water solution taken up through plants' roots or, in animals, by drinking; (2) from an atmospheric reservoir as an organism breathes [this is often the case with gaseous elements and compounds, e.g., oxygen, nitrogen, or carbon dioxide (CO_2)]; or quite commonly (3) from eating the body or wastes of another organism (Figure 2-2).

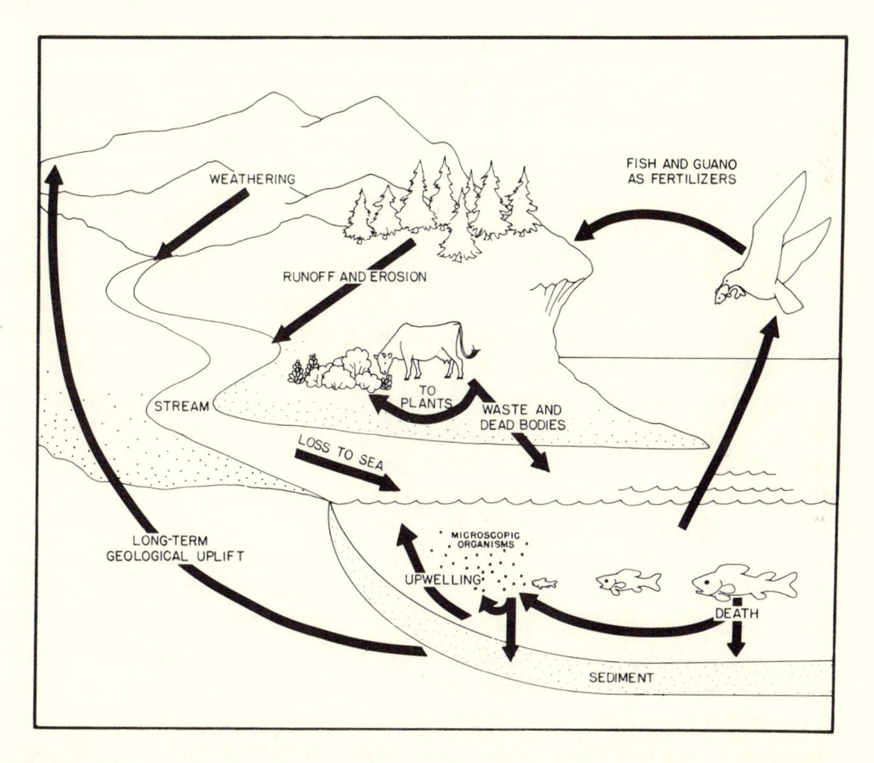

FIGURE 2-3. The cycling of phosphorus in the biosphere. The element is taken up by plants and animals in nonorganic form in water solution and is incorporated into the bodies of animals eating the plants or eating other animals and wastes of animals and plants. Some phosphorus from animal and plant sources is returned to inorganic form in water solution or through sedimentation. Phosphorus is lost from the cycle if it becomes inaccessible to organisms.

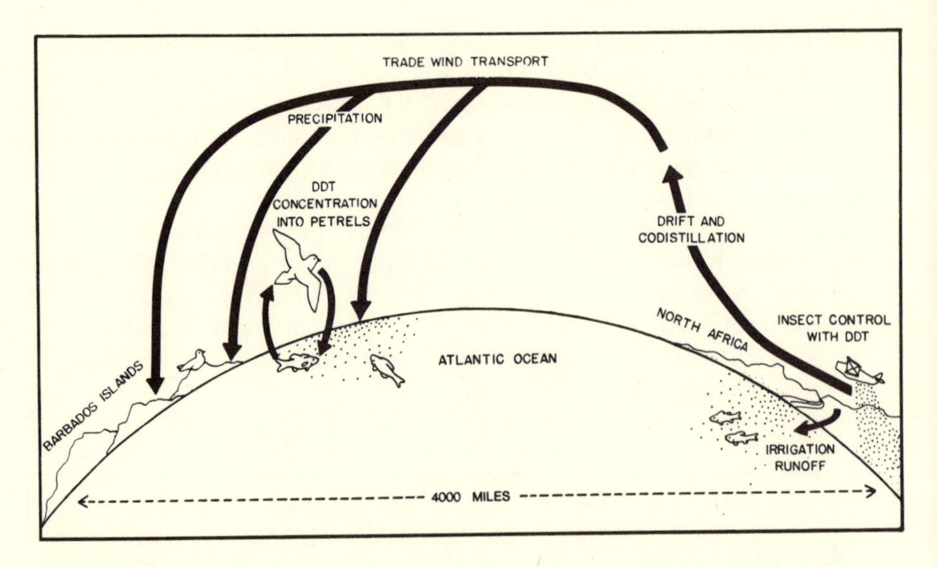

FIGURE 2-4. An example of the transport of a toxic material, DDT, in the biosphere. A portion of the DDT sprayed for locust control in North Africa is carried through the atmosphere and may be washed down in precipitation thousands of miles away, along the coast of the Americas. It may be distributed locally in irrigation runoff as well (redrawn, with additions, from Whittaker, 1975).

The amount of these elements in the biosphere is finite. Malfunction and noncompletion of the complex cycling process that makes these elements available would not cause their disappearance from the earth's surface. But breaks or inefficiencies in a cycle could severely limit the availability of these essential materials to the organisms that depend so clearly on their presence. An example will illustrate.

One of the basic building blocks of life is phosphorus. Every living organism requires phosphorus for the metabolic processes of energy storage and release. Farmers recognize this requirement when they fertilize with materials containing phosphorus.

The earth's principal sources of phosphorus are rocks and other deposits formed during past geological ages. In such locations and forms it is not readily available to most living organisms. However, phosphorus from these deposits is slowly eroded away and washed down to lakes, oceans, and other bodies of water in solution. As such it is taken up by marine organisms and plants growing along these phosphorus-rich areas. Fish, birds, and other animals that in turn eat these organisms may transport the phosphorus to less well-endowed areas in their waste materials or through decomposition of their bodies after they die. This phosphorus will be utilized by plants and animals in the new area until it is carried elsewhere, either to other habitats or to the ocean depths, where it is out of reach to living organisms (Figure 2-3).

The often rapid spread and the extent of transport of materials between the physical environment and its interacting organisms has been dramatically illustrated in recent years by the discovery of toxic concentrations of DDT and other biological poisons in organisms far from their application site (Figure 2-4). These materials, like the essential elements, travel along both biological (i.e., in organisms) and physical (e.g., with wind and water) pathways.

The other basic integrating factor in the biosphere, the flow of energy, is not a cycling system but a one-way transfer of energy through one organism to another (via food chains, as they eat each other) until it is dissipated (Figure 2-5). This is possible because the source of energy (the sun), unlike the reservoirs of material, is outside the biosphere and for all practical considerations apparently infinite.

Although some of the sun's energy is utilized directly by the biosphere's organisms as heat, most organisms get their energy secondhand. Energy is stored in the sugar molecule during the photosynthetic process by green plants and some bacteria. These sugars can later be broken down by the plants themselves or other organisms to release the chemically stored energy for use in survival and growth.

The photosynthetic reaction in green plants can be described chemically as follows:

$$\text{Carbon dioxide} + \text{Water} + \text{Energy} \xrightarrow[\text{of chlorophyll}]{\text{In the presence}} \text{Glucose} + \text{Oxygen} + \text{Water}$$

$$6CO_2 + 12H_2O + \text{Sun's energy} \xrightarrow{\text{Chlorophyll}} 6(CH_2O) + 6O_2 + 6H_2O$$

Briefly, in the chlorophyll-containing parts of the plants (the chloroplasts), the sun's energy is used to split water into its hydrogen and oxygen components, resulting in a release of energy. This energy is used to combine the "loose" hydrogen molecules with carbon dioxide from the air to form sugars such as glucose (Figure 2-6).

There are several prerequisites for photosynthesis. First, of course, is the sun's energy, not only for the light energy that is used to split the water molecule but also for heat. Like other metabolic processes, photosynthesis ceases at high- and low-temperature extremes. Second, water

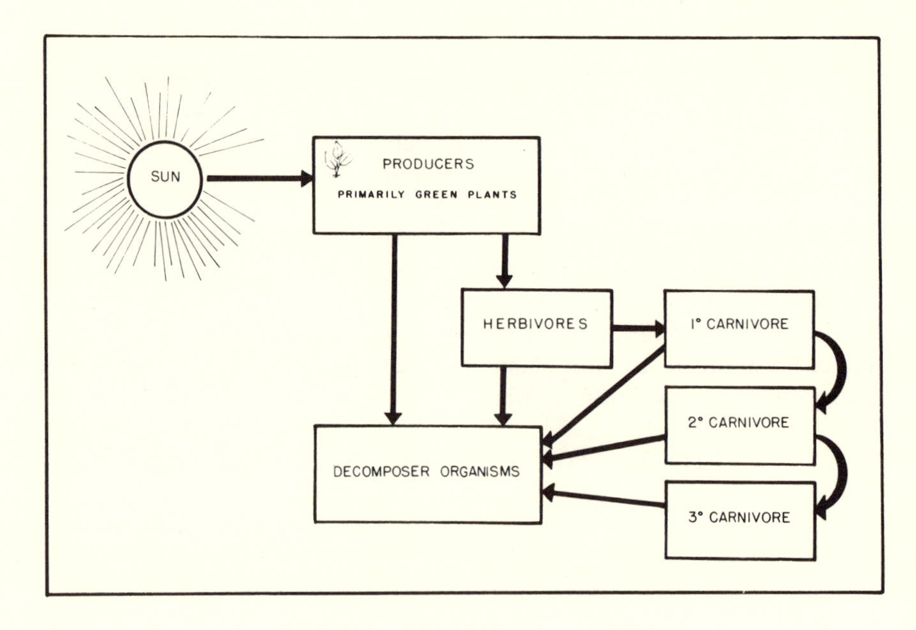

FIGURE 2-7. Elements of a food chain.

The energy released in respiration is used by plants and animals to carry out the metabolic functions necessary to maintain life growth.

Not all the energy bound in photosynthesis is utilized by the plant or even the animals and microorganisms that feed directly on it. In fact, enough energy is passed on to the carnivorous animal, which eats each of these plant-feeders (or herbivores), to feed a succession of carnivorous (or animal-feeding) animals and finally to power the activities of the decomposer organisms, which consume the remaining dead plant and animal material. Such a relationship of organisms successively feeding on each other is known as a *food chain* (Figure 2-7).

At the beginning of any food chain is the *autotroph* (self feeder) or *primary producer*, the organism that transforms energy from a nonorganismic (abiotic) source (e.g., by photosynthesis or, rarely, by chemosynthesis) into organic molecules. The remaining organisms along the food chain are known as *heterotrophs* and depend on the energy-rich organic molecules produced by autotrophs and other organisms for their energy supply.

The first heterotroph on the food chain is the *herbivore*. Following the herbivore are a series of *carnivores* all feeding on living animals. The

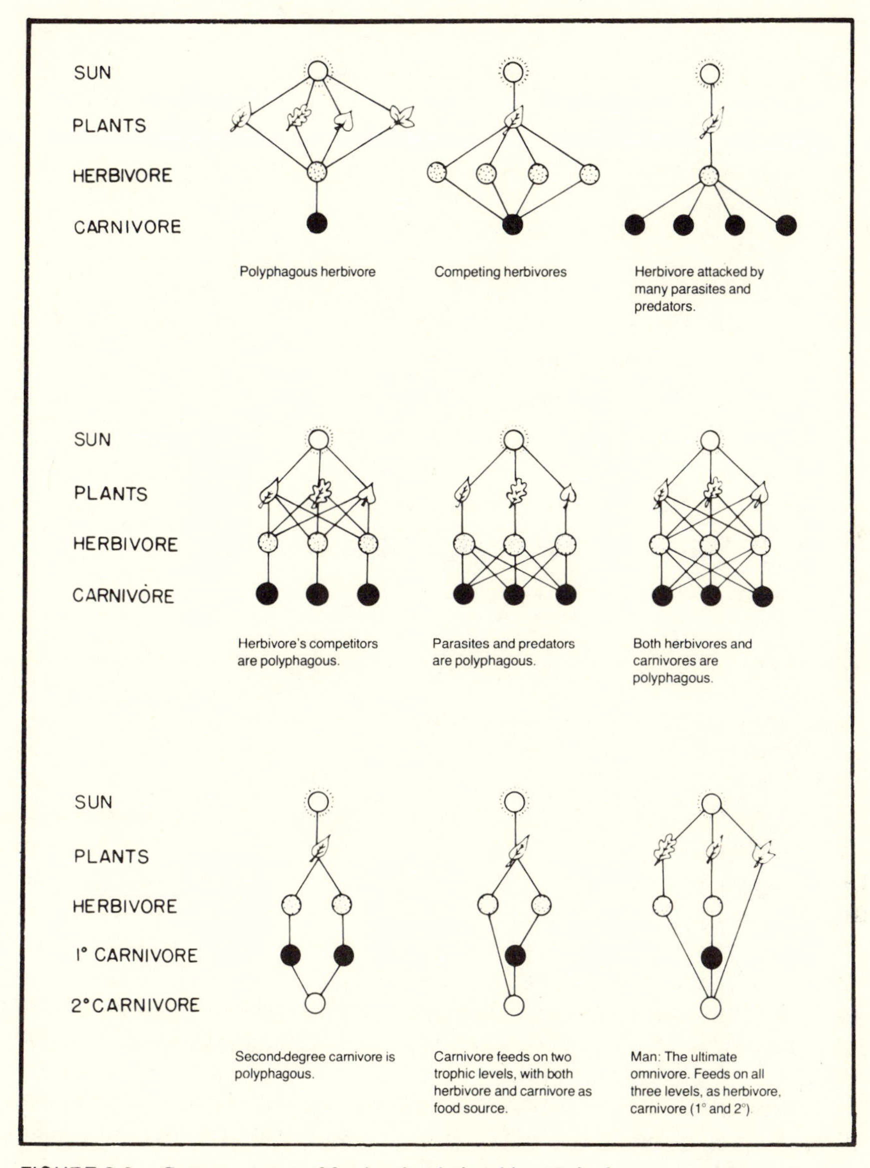

FIGURE 2-8. Common types of food web relationships. *Polyphagous organisms* eat more than one kind of food (as opposed to *monophagous organisms*, which will feed only on one species of plant or animal). Predators and parasites are two types of carnivores.

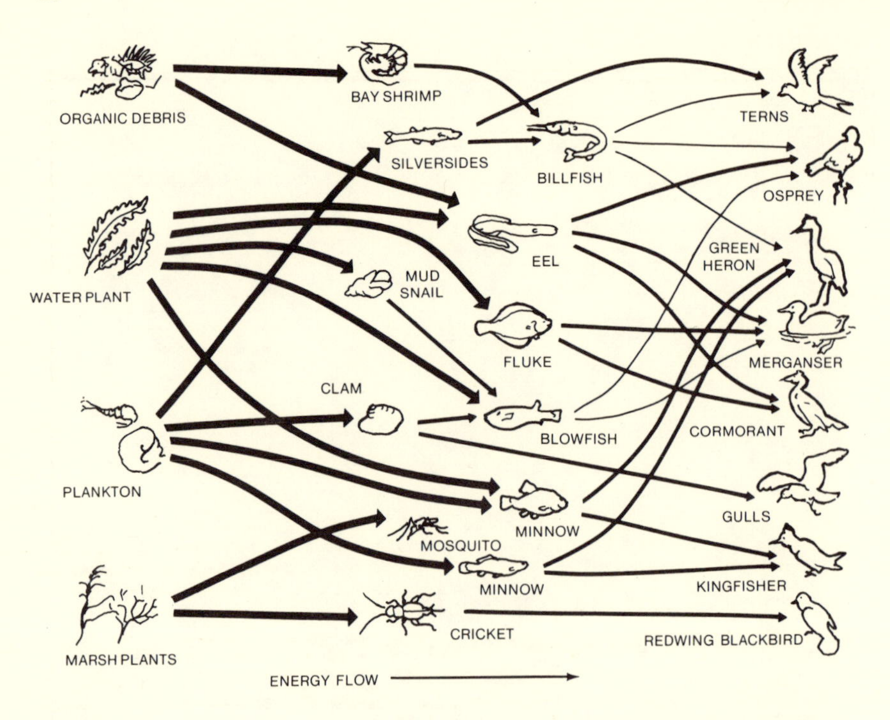

FIGURE 2-9. Portion of a food web in a Long Island, New York, estuary. Arrows indicate flow of energy. Note how many organisms (e.g., blowfish, terns) are feeding at more than one trophic level at once; this gives these relationships a "weblike" character (after Woodwell, 1967).

primary (1°) carnivore consumes the herbivore; it, in turn, is devoured by a secondary (2°) carnivore; the tertiary (3°) carnivore feeds on the secondary, and so on. The last heterotroph on any food chain is a *decomposer organism*. These decomposer or reducer organisms are often called the "enzymes of the biosphere" because of their importance in breaking down dead plant and animal material into nutrients again usable by autotrophic organisms.

Food chains describe the direction of energy flow. In reality every plant or animal has not one but several different organisms that may feed on it, regardless of its trophic (feeding) level. Additionally, many organisms feed on a variety of foods and thus may place themselves on two or more trophic levels at once. In other words, the energy locked in an organic molecule by a given autotroph may be passed through a variety of pathways (food chains) before it is exhausted. A *food web* describes these pathway possibilities. Figure 2-8 shows common types of food web relationships. Figure 2-9 shows a partial food web in a marine estuary. Even though this diagram represents only a small fraction of the biota in such an ecosystem and doesn't follow all the energy captured by the plants at the start of the food chain through to its final release, it does make clear the degree of food web complexity in natural ecosystems.

Figure 2-10 explains why toxic substances are likely to concentrate at the highest trophic levels on the food chain, and why it is in these organisms (like birds and fish) that we often see the effects of pesticide overuse first. Figure 2-11 shows the build-up of such a toxic substance, the insecticide DDT, in the Lake Michigan food web. Here, as could be

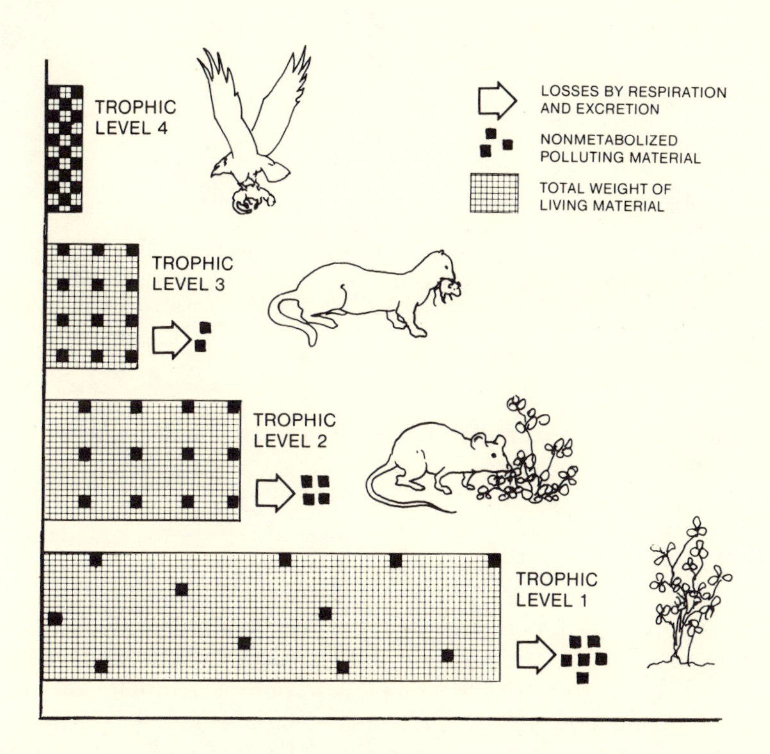

FIGURE 2-10. Diagrammatic representation showing why toxic substances are most likely to accumulate in animals at the highest trophic levels on the food chain. Large quantities of food material are taken in by each organism, but these are lost through respiration and excretion. If the chemical is not metabolized by the cells and remains in the animal's body tissues, the result is an ever-increasing concentration of the material in the animal as more food is taken in.

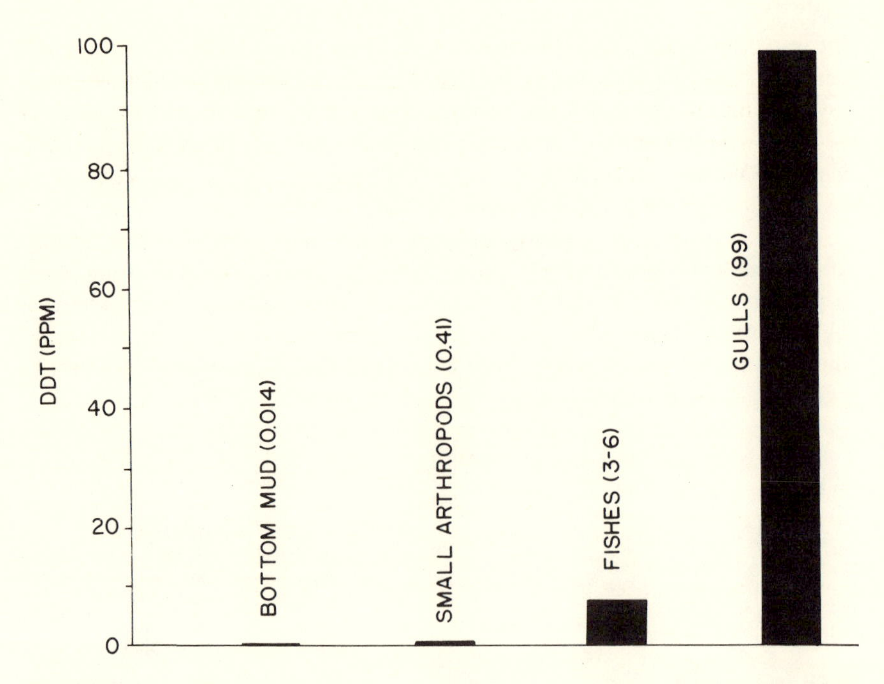

FIGURE 2-11. DDT concentration in a Lake Michigan food chain. DDT load in gulls is about 240 times that of small arthropods. In 1969 Coho salmon from Lake Michigan were found to contain DDT at 4 to 6 times the tolerance standard for human consumption (redrawn from Berry *et al.*, 1974).

predicted, the concentration of DDT is greatest at the highest trophic level—the fish-eating gulls.

COMMUNITIES AND ECOSYSTEMS

Within the biosphere are groups of interacting and interdependent populations of plants, animals, and microorganisms that share a common environment and form a fairly complete trophic structure for the capture and release of energy and the cycling of material within the environment; these groups are known as *communities*. Examples of communities include all the organisms in a salt marsh, forest, prairie, or the organisms on, in, and under a decaying log.

There are no size constraints as such on the concept of "community." In fact, we may describe communities within communities. For example, the organisms in, on, and alongside a small pond within a forest may be seen as a community with a functioning trophic structure of primary producers, consumers, and reducers of its own; in addition, the pond organisms may be included as part of the larger surrounding forest community unit. In other words, communities are merely human constructs through which we may view and better understand the biological and ecological interrelationships between organisms.

An *ecosystem* is a community and its nonliving support system. For example, a forest community consists of all the forest-inhabiting organisms; its corresponding forest ecosystem includes, along with all these organisms, the soil, rocks, water, air, and other physical components of the forest landscape. The ecosystem concept is probably more useful than the community subdivision because, like the biosphere concept, it stresses the constant interaction and interdependence between the living and nonliving elements of any environment.

The evolution of ecosystems reflects this interdependence. The physical environment (for example, temperature extremes, water and mineral availability, altitude) sets the limits to the type and quantity of plant and animal life that can successfully survive and reproduce. On the other hand, there is a constant modification of the nonliving environment by its resident organisms.

The progressive development or "successful trends" of the plants, animals, and physical character of an ecosystem as it evolves through time follow a pattern of increasing complexity of growth habit, physical

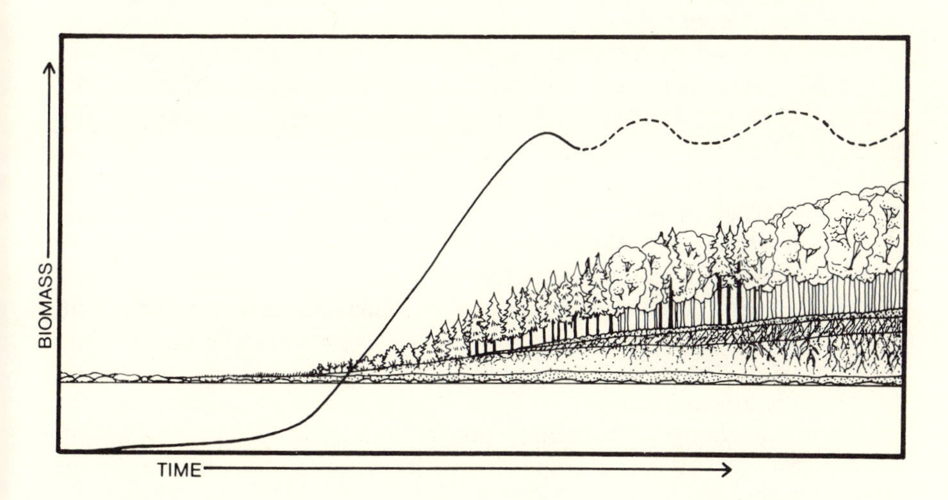

FIGURE 2-12. Succession from a bare surface of rock to a permanent forest ecosystem, showing the increase in biomass (the living component of the ecosystem). Note also the accumulation of organic debris and larger soil layers. Concurrent with the increase in size and numbers of representative plant species is a similar increase in animal size, number, and diversity.

diversity, and species interaction. For example, this can be illustrated by following the evolution of a bare surface of rock from its first exposure by a landslide to the development of a permanent forest ecosystem (Figure 2-12).

Long-buried, the new landslide-exposed surface of rock is at first barren and lifeless. Lichens, blown in with the wind or rain, are usually the first organisms to settle successfully in this unlikely habitat. Slowly, through the symbiotic functioning of its component parts, the lichen eats away at the rock. The rock-clutching fungus contributes a capacity for holding water, nutrient uptake, and chemical defense; its partners, the algae, are able to photosynthesize. The wastes and bodies of generations of such lichens result in the accumulation of a small amount of organic debris; this, along with small chips of rock, forms a substrate that can better retain water and provide an anchoring place for the roots of the first higher plants. Early representatives, whose spores or seeds are blown in or brought by migrating birds and other animals, are often mosses, followed by grasses; each produces more and more organic debris, improves water and nutrient retention, and attracts and supports an increasingly diverse macro- and microorganismal fauna and flora. Later, shrubs arrive, then trees; these newcomers often eventually shade out the previous inhabitants. As the rocks are worn down and buried under a larger accumulation of rich organic soil, new and larger species of trees and wildlife move in and, after many years, begin to form a potentially permanent forest community.

In such a successional development we can recognize several trends (after Whittaker, 1975) taking place in the evolution of both the biotic and abiotic components of the ecosystem:

1. There is a progressive development of the soil through wearing down of physical (e.g., rock) components and addition of organic matter. Increasing depth, increasing organic content, and increasing differentiation of soil layers characterize this development.
2. The plant community becomes taller and more massive and shows more differentiation and variability as time goes on. Similarly, the animal community grows in size and becomes more diverse in habit.
3. The quantity of organic nutrients available to organisms increases. A larger proportion of these nutrients is held in living tissues as the community evolves.
4. Production of organic material (that is, photosynthesis) per unit area goes up as plants become larger, longer-lived, and more diverse.
5. As the living components of the ecosystem become larger and

more dense, they begin to predominate over the physical environment in the determination of future evolution.

6. Species diversity, in general, increases from a few species in early communities to many species in communities at later successional stages.
7. Species come and go during the course of succession. In the plant community, especially, these changes tend to occur more slowly in later communities as smaller and shorter-lived species are replaced by larger and longer-lived ones.
8. The increase in species and diversification of habit of these species makes for a more complex food web and a more stable ecosystem. When there are a variety of food chain possibilities, the ecosystem will be less effected by loss of a few species than at earlier stages.

Over the short run, natural ecosystems maintain a fairly constant character. This is due primarily to constraints set by the physical envi-

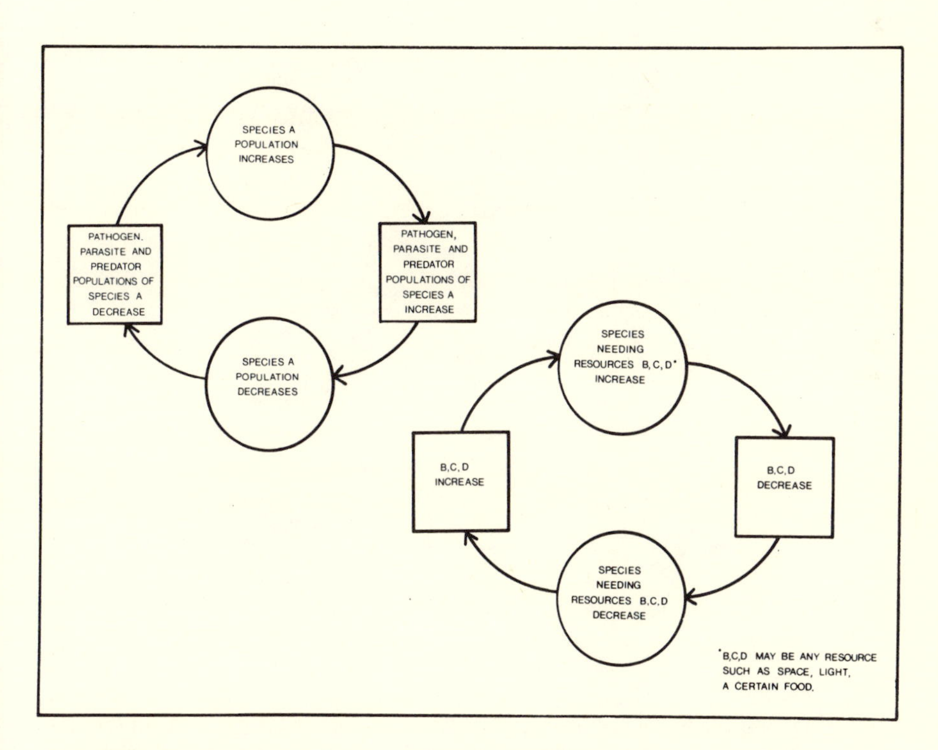

FIGURE 2-13. Feedback mechanisms regulating population numbers. Top, predation; bottom, competition.

ronment together with feedback mechanisms operating on both intra- and interspecies levels. One of these mechanisms is predation (Figure 2-13, top). For instance, when the population of one species increases, more food becomes available to all the organisms which feed on it (its natural enemies). As a result, populations of these natural enemies may also become larger. And the increased numbers of these pathogens, parasites, and predators (or natural enemies) soon lower the population of their food source organism. Another major feedback mechanism is competition (Figure 2-13, bottom). As the population of a species (and others requiring the same resources) gets larger, it becomes more and more difficult to acquire the food, water, and space necessary for growth, reproduction, and survival. Consequently, above certain population levels the reproductive rate is lowered and the mortality rate increases. In these ways population numbers stay within certain upper and lower limits—often called the *characteristic abundance* in a given ecosystem. E. P. Odum (1971) illustrates characteristic abundance well: "a forest may have 10 birds per hectare and 2000 arthropods per square meter, but never the reverse."

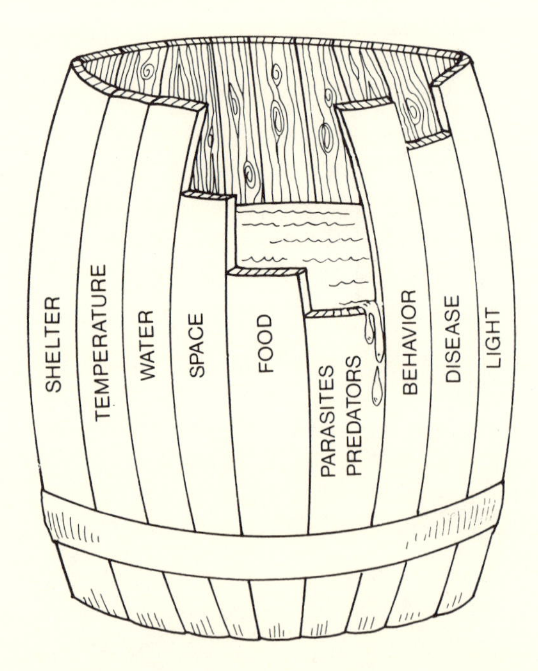

FIGURE 2-14. A barrel diagram that illustrates the concept of *limiting factors*. At various times any of a number of factors may limit the number of individuals of a species able to survive in a prescribed area. In this diagram some of these factors are represented by the barrel's staves. At this time parasites and predators are limiting the population size—that's where the water is leaking out of the barrel!

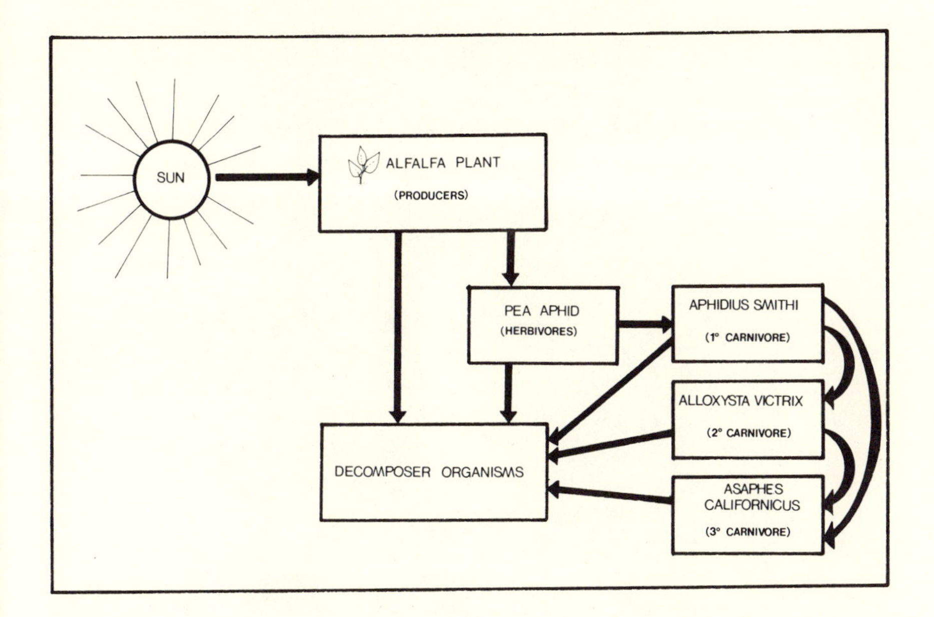

FIGURE 2-15. A food chain in the alfalfa ecosystem. The pea aphid is a common pest in alfalfa. Note how *Asaphes californicus* may act either as a secondary (2°) or tertiary (3°) carnivore (after van den Bosch and Messenger, 1973).

The factors that commonly keep populations below certain numbers may be called *limiting factors* (Figure 2-14). For instance, plant size and population numbers are often limited by availability of water, nutrients, light, and growing space. The key factor in maintaining numbers of fish in a given area is very often food. On the other hand, plant-eating insects in natural ecosystems rarely eat all the available food but are instead kept at their characteristic abundance level by the appetites of the various carnivores and pathogenic organisms (natural enemies) that feed on them. Territorial behavior limits many bird populations. Surplus birds will not be able to find the nesting sites necessary for reproduction; in this way, the growth of unsupportable populations is prevented before new individuals are ever born.

HUMAN-MANAGED ECOSYSTEMS

Agricultural, silvicultural, and other human-managed environments are ecosystems too. Each has a trophic structure of organisms that captures and releases energy and cycles material through the system. Figures

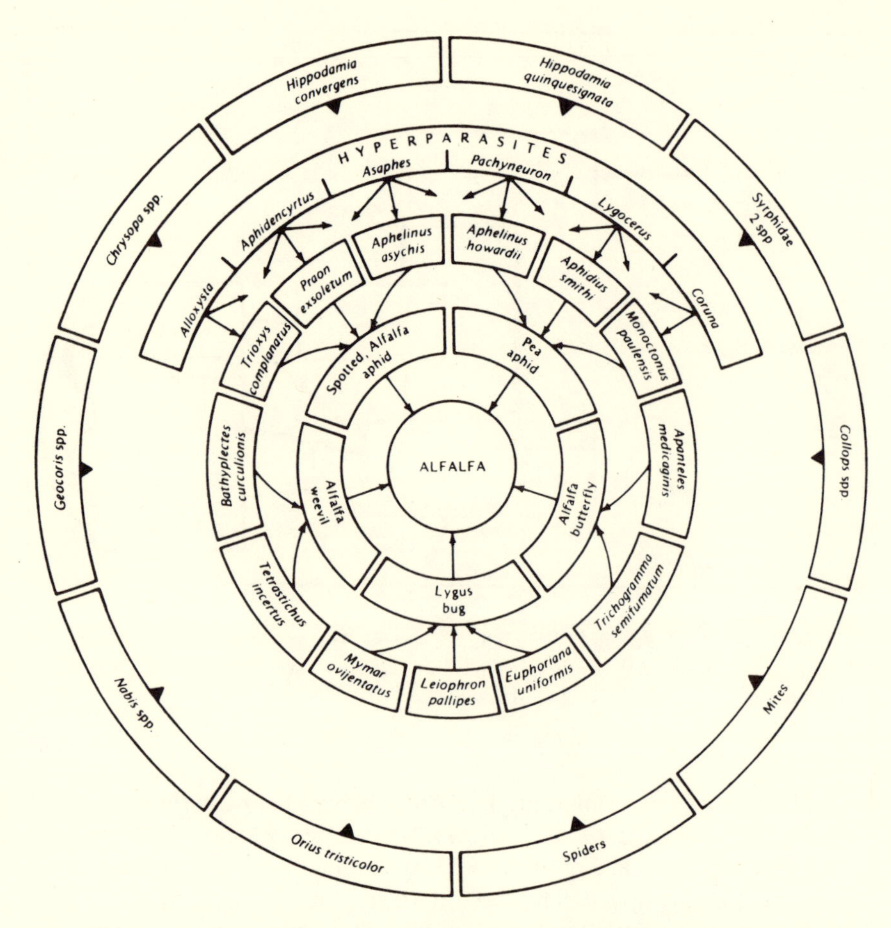

FIGURE 2-16. Partial food web showing common insects in the alfalfa agroecosystem. The insects listed in the ring surrounding the alfalfa circle are phytophagous pests. The next ring contains the main primary parasites of these pests. The third ring includes secondary and tertiary parasites (hyperparasites), which attack some of the parasites in the second ring. The outer ring includes predators, some of which (*Hippodamia* spp., for example) attack only aphids and others, general predators, that attack most small-bodied inhabitants of the alfalfa plant (from van den Bosch and Messenger, 1973).

2-15 and 2-16 show a food chain and a partial food web in the alfalfa agroecosystem.

However, because people manipulate the ecosystem to increase their own benefits there is a much larger importation of energy and materials from sources outside the ecosystem than in nonmanaged situations (Figure 2-17). For instance, in an agro-ecosystem, importations from outside the ecosystem might include

1. Fertilizers, insecticides, herbicides, fungicides, molluscides, nematicides, and other agrichemicals (like growth regulators, defoliators)
2. Crop seeds or seedlings
3. Fossil fuels, human labor, and other energy from outside the ecosystem
4. Water, if the field is irrigated

Because the manager usually removes a product and undesirable materials, much more material and energy leave the agricultural ecosys-

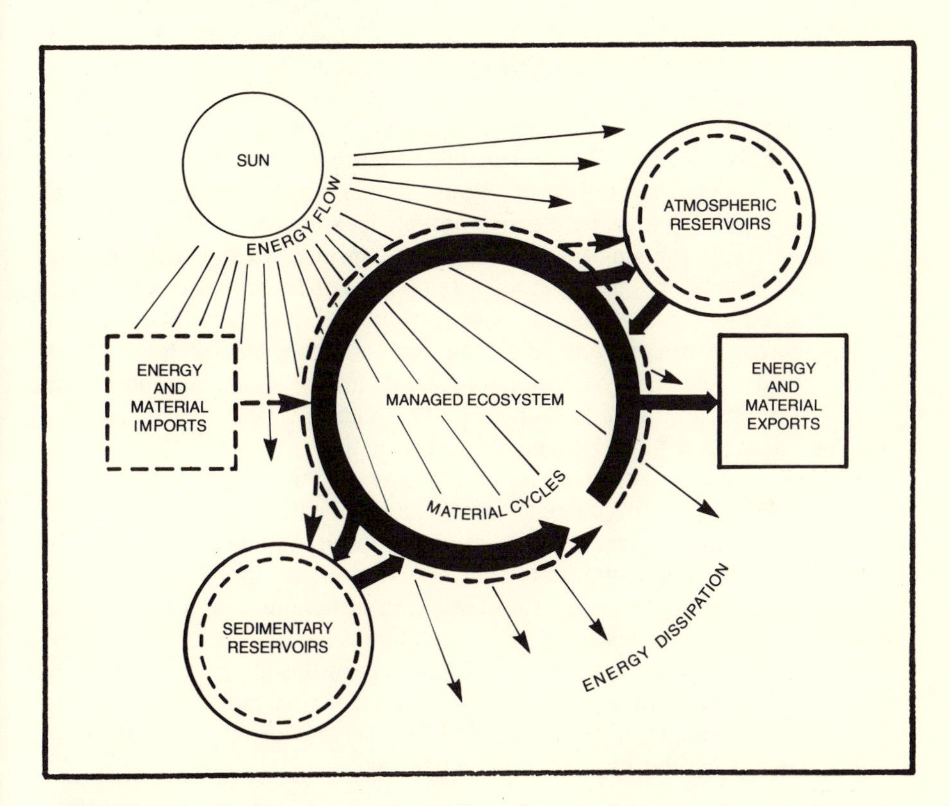

FIGURE 2-17. Cycling of materials and flow of energy from natural sources is greatly supplemented in managed ecosystems by importations of materials and energy from outside the ecosystem and is also significantly depleted by exportations beyond the boundaries of the ecosystem.

tem than in a natural situation. Thus, in the agro-ecosystem, the following might be exported:

1. The harvested crop
2. Weeds
3. Plant prunings
4. Extra seedlings (through thinning operations)
5. Nutrients (exported with product and wastes or leached out in irrigation).

Compared to natural ecosystems, diversity is decreased on all levels in agro-ecosystems. Genetic diversity, for instance, has been drastically reduced in crop plants (and many ornamentals) to increase yield, attractiveness, and resistance to certain pests—often causing a consequent reduction of natural resistance to other pests and stresses in the physical environment. Species diversity is lost; a monoculture, with its near exclusion of alternate food and shelter sources, produces a drop in species diversity at all successive trophic levels. The variety of plant growth forms (vertical, horizontal, and structural diversity) is low in agro-ecosystems; same-aged plants are similar in size and structure and provide no diversity in shading and shelter for other organisms. The even spacing of plants in most agro-ecosystems results in a notable lack of spatial diversity, thus making it easier for small herbivores (e.g., insects) to find their hosts. Agro-ecosystems are primarily of a temporary nature, and changes are sudden, frequent, and simultaneous; consequently, there is little variety in age of crop plants, weeds, and organisms at higher trophic levels. The diversity, irregularities, and deficiencies of the physical environment in the agro-ecosystem have been largely eliminated by land clearing, burning, plowing, leveling, fertilization, and irrigation.

In many ways, with this loss of diversity on nearly every level, managed ecosystems are often like the early stages of natural succession. Like early successional stages, there is limited differentiation in age and growth form within the plant community. The pool of available inorganic nutrients in these managed systems is often small and from nonliving sources (i.e., fertilizer). Also, species diversity is relatively low, resulting in shorter food chains and more simple food webs than in the more stable natural ecosystems. Even more so than early successional stages, managed ecosystems are unstable and often could not survive without constant input and upkeep provided by their human caretakers.

CHAPTER 3

What is a Pest?

We share this planet with the million or more other species of organisms that inhabit the biosphere. This mass of organisms goes on reproducing; dying; taking up space; taking in water, air, and essential nutrients; depositing waste materials; and, most important, maintaining a delicate hierarchical balance of species, constantly eating one another. It goes on so quietly and unobtrusively that human beings are rarely conscious of this vast exchange taking place before their very eyes and beneath their very feet.

Which of these organisms are considered pests? Why are they considered pests? And when do they become pests? These seemingly simple questions are, in fact, the very heart and soul of sound integrated pest management. Two situations commonly predicate failure of a pest control attempt: (1) without valid reason, an organism is labeled a "pest" and decimated, and the surrounding ecosystem is thrown out of balance, releasing a whole spectrum of new "pest" problems; or (2) the pest manager simply tries to kill the pest (a temporary solution) instead of discovering the reason why the organism has become a problem and changing that situation (a permanent solution).

Broadly speaking, a *pest* is an organism that reduces the availability, quality, or value of some human resource. This resource may be a plant or animal grown for food, fiber, or pleasure (e.g., pets, or plants in homes, backyards, and park and recreation areas). The resource may also be a person's health, well-being, or peace of mind—one or all of which may be threatened from time to time by allergy-inducing or otherwise bothersome plants, disease-vectoring organisms, or biting, stinging, and nuisance-type animals such as gnats, lovebugs, and cockroaches.

Organisms that have become "pests" are not limited to any class or phylum. They are as varied as the habits that make them undesirable. Expectedly, insects, the most diverse and numerous class of animals, are frequent pests. Certain mite, tick, nematode, mollusc, and other invertebrate species are known to become pests. Vertebrates, including rodents, deer, coyotes, and birds, may become serious pests in some situations. Microorganisms (bacteria, fungi, protozoa, rickettsiae, viruses, mycoplasmae), particularly those that are pathogenic to important plants and animals, cause many problems. Weeds—plants in places where they are not wanted—comprise another category of common pestiferous organisms.

Most organisms, of course, are not pests. Many are beneficial—e.g., the organisms that we eat and enjoy, or the organisms that destroy unwanted organisms: the natural enemies of our pests. But the overwhelming majority of organisms in the biosphere are never classified as pest or beneficial. We usually don't concern ourselves with these plants, animals, or microorganisms because we are not aware of their playing a significant role in our lives. We must change that attitude, however, as all are important components of the food chains that maintain the natural balance of the biosphere.

It is important to realize, too, that "pests" are defined according to human needs and values—and contemporary needs and values at that. If people ever decide to inhabit only concrete, glass, and steel buildings, termites will cease to be pests and will become appreciated as beneficials for their role as important decomposers in the forest ecosystem. If Americans ever acquire a taste for insect larvae (an excellent source of protein, incidently) as have certain people in Africa and Australia, we might come to view plant-consuming locusts or log-boring beetles a bit differently as they munch or gnaw unhindered under the now-protective eye of the "bug" farmer—joining cattle, sheep, and swine in the ranks of valued livestock.

These examples may seem a bit farfetched, but they do point out that the designation of an organism as a pest depends on its situation rather than its species—or even its trophic position. For instance, a comparison of two cotton plants, one growing wild in its native Central America and the other in a highly cultivated monoculture in Texas, would show that while both have many of the same insect species feeding on them—aphids, weevils, caterpillars—their importance in each situation would be different. In the cultivated Texas cotton fields, these insects may sometimes be considered major pests if they are numerous enough to shrink the harvestable yield; in the wild the same insects go almost unnoticed as minor elements in a larger ecosystem.

ECONOMIC LOSS IS PRIMARILY A FUNCTION OF POPULATION DENSITY

One, ten, or even a thousand leaf-feeding caterpillars on a peach tree may not create a pest problem—but 10,000 caterpillars on the tree may well be a problem. Pest problems are most frequently the result of a species population becoming and remaining, for some reason, more dense than normal.

Even in highly dynamic (unstable) field crop situations, population densities of an insect species stay fairly stable over a long period of time (see discussion of *characteristic abundance* in Chapter 2) and can be said to fluctuate around a mean level, which entomologists call the *equilibrium position*. The maintenance of this equilibrium position (or characteristic abundance) is due primarily to constraints set by the physical environment together with feedback mechanisms operating on both intra- and inter-species levels. These mechanisms—such as predation, parasitism, intraspecific competition for food or available nesting sites, and territori-

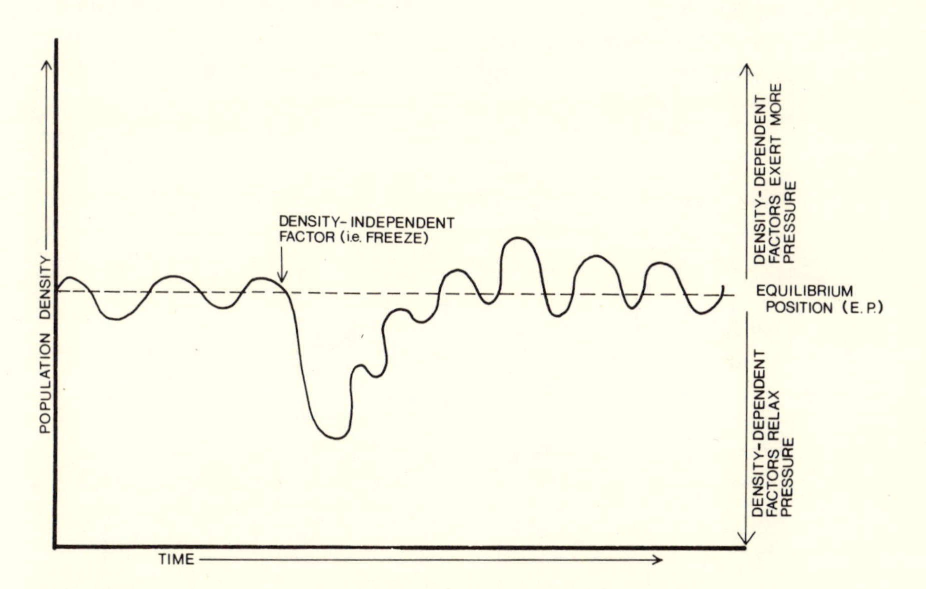

FIGURE 3-1. Influence of density-independent and density-dependent regulating factors on the equilibrium position of a population. Note how density-dependent factors exert more pressure on the population as it gets more dense. Whereas density-independent factors can cause sudden, drastic changes in population numbers, it is the density-dependent factors that are responsible for maintaining the long-term characteristic abundance level of the population.

ality—characteristically exert much heavier pressures when populations are high and have diminished influence when populations are low; they are thus called *density-dependent* factors (Figure 3-1).

Other components of natural control (Figure 3-2), *density-independent factors*, limit different-sized populations to degrees independent of their density and thus are relatively not as important in maintaining the equilibrium level. Such a factor may be the killing of many individuals in the population by a storm or early freeze. Since the freeze might wipe out 50% of the population, whether it is composed of 60 or 600 individuals, its effect can be called density-independent. Yet, as we can see in Figure 3-1, a population reduced to a level where density-dependent factors are relaxed will rapidly surge back up to the equilibrium level—or its characteristic abundance.

In a managed ecosystem there are various ways to manipulate the equilibrium position of a pest. Although sometimes initially expensive, these are usually the most satisfactory and cheapest ways to control pests in the long run because they permanently reduce the number of pests that can occupy a given area. One of these ways is to add another permanent density-dependent factor: for instance, the introduction and establishment in the ecosystem of a new natural enemy (Figure 3-3). This method, known as *biological control*, has been utilized effectively scores of times for the

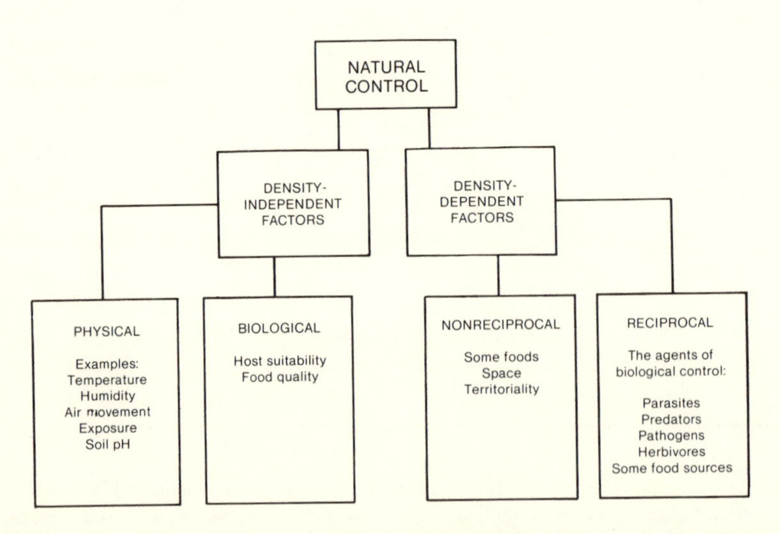

FIGURE 3-2. The components of natural control. Reciprocal density-dependent factors are those in which feedback mechanisms (see Chapter 2, Figure 2-13) play a role in determining the supply of the factor as well as the population density of the organism in question. The quantity of nonreciprocal density-dependent factors is not affected by pest population levels (from van den Bosch and Messenger, 1973).

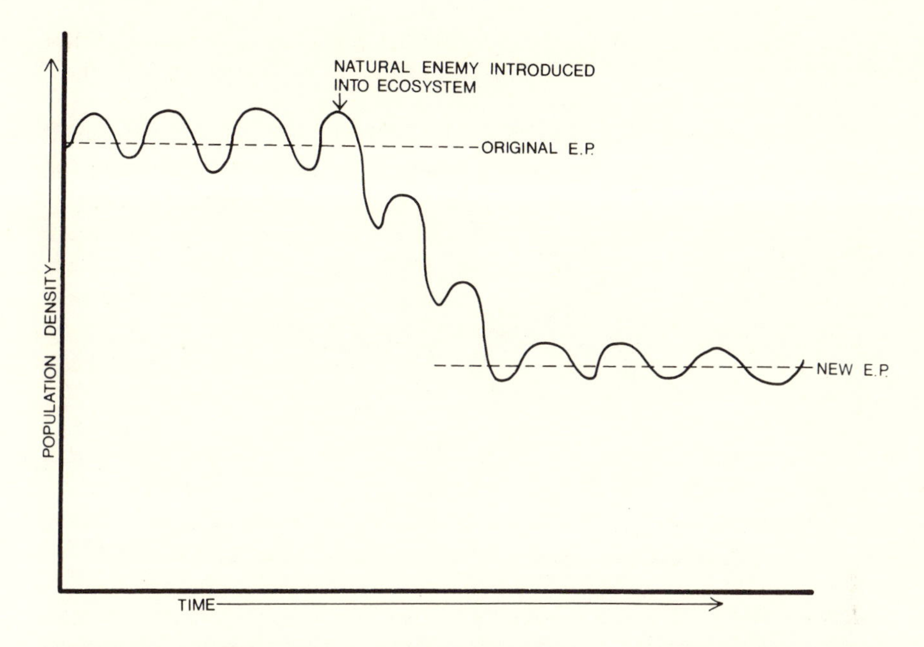

FIGURE 3-3. Lowering of the equilibrium position by the introduction of a new natural enemy. The same effect can be obtained by changing the physical environment so that fewer individuals can survive.

control of both insect pests and weeds. Examples of biological control will be discussed in later chapters.

Another way to lower a pest's equilibrium position is to change its physical environment permanently so that fewer individuals can survive. This is an effective way to control rats—if the number of possible nesting sites is reduced, fewer rats can breed and the equilibrium level is reduced (as in Figure 3-3). Other environmental manipulations that may lower the equilibrium position are the permanent removal of food and water sources, overwintering sites, breeding areas, and refuges.

Pest management practices can inadvertently raise the equilibrium position as well. An insecticide application for control of a pest sometimes does a more effective job of knocking out a natural enemy; if the insecticide is regularly applied, the result may be the permanent impeding of a density-dependent factor—the natural enemy—and a higher equilibrium level for the pest. The equilibrium level may also be raised when poor management of an ecosystem provides new breeding places for pests—e.g., an increase in garbage for flies and stagnant pools of water for mosquitoes (Figure 3-4).

How large must a population get before it can properly be called a

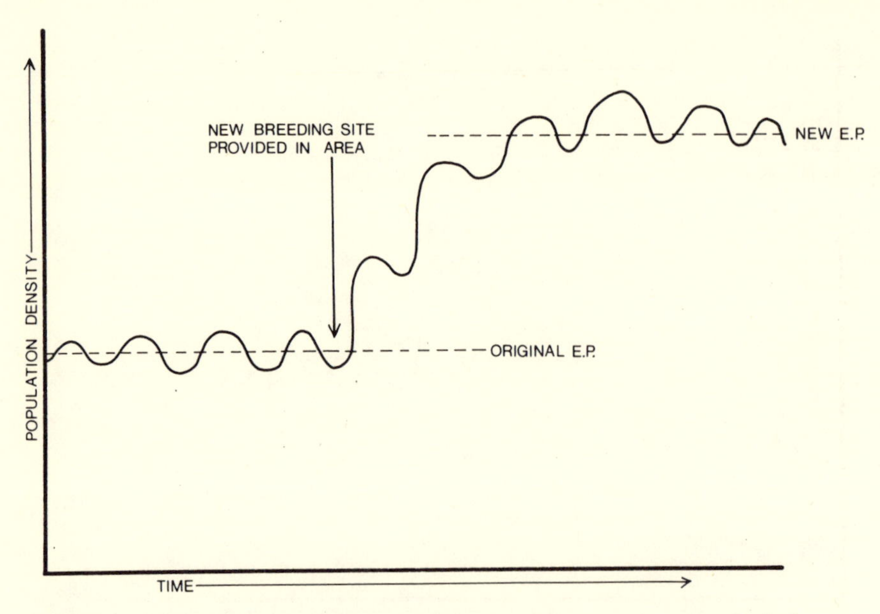

FIGURE 3-4. Raising of the equilibrium position by adding a new breeding site in the area. The removal (e.g., owing to repeated insecticide use) of an important natural enemy species from the ecosystem would have the same effect.

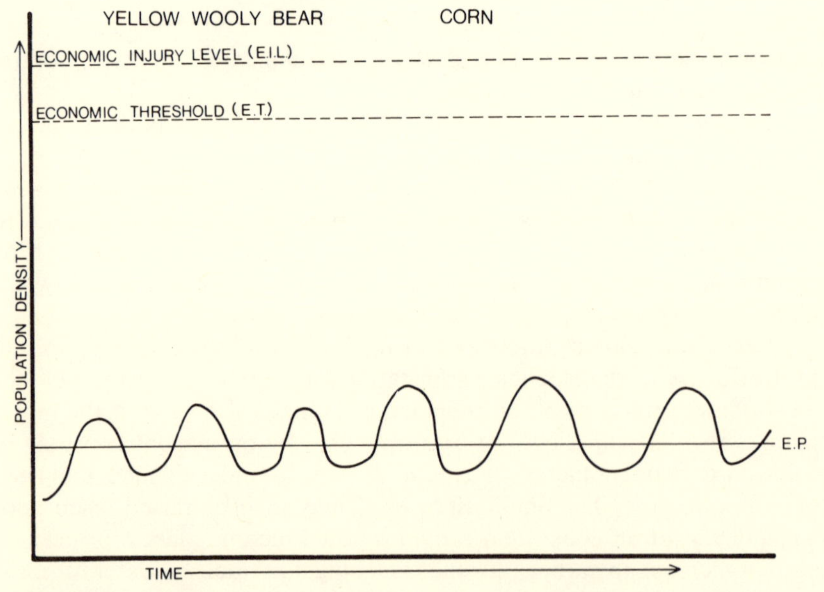

FIGURE 3-5. Equilibrium level is well below economic injury level. Such an insect is never a pest; control action is never needed (after Luckmann and Metcalf, 1975).

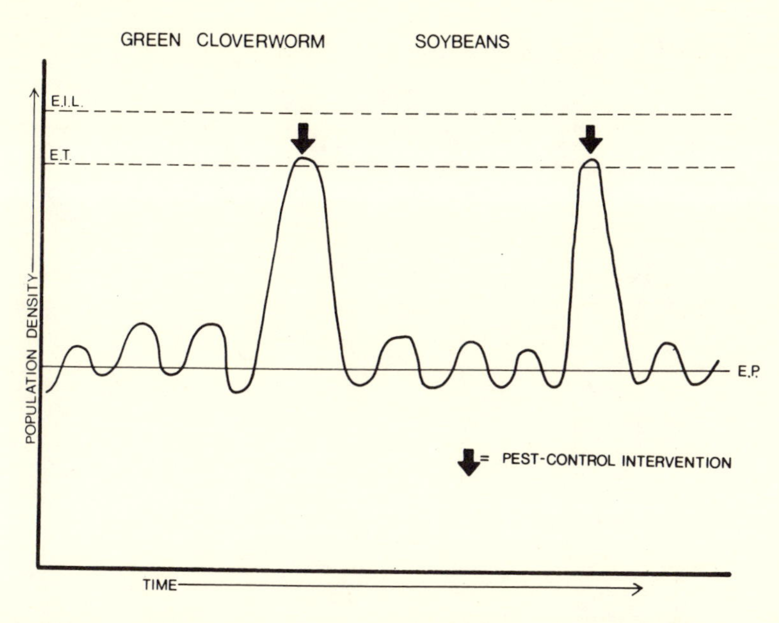

FIGURE 3-6. Equilibrium level is well below economic injury level, but for various reasons the population density sometimes reaches the economic injury level. This is an occasional pest; occasional control action is needed (after Luckmann and Metcalf, 1975).

pest? Entomologists have developed the concept of *economic injury levels* to deal with this question. Economic injury levels indicate the pest densities (numbers of pests per unit area) at which artificial control measures are economically justified. In other words, at these levels the cost of control is less than the loss the farmer, forester, or other resource producer would suffer if control action were not taken. In urban areas, recreation areas, and other situations where the goal of pest management is the provision of an enjoyable environment rather than an economic return, *aesthetic injury levels* indicate the densities at which pest damage is severe enough to justify the cost of control. Accurate determination and careful use of these levels is essential to the expansion of good pest management and the maintenance of environmental quality.

If the economic injury level for an insect species is well above its equilibrium position, the insect is not likely to become a pest very often (Figures 3-5 and 3-6). However, insect populations with average densities close to their economic injury levels are more likely to be familiar pests (Figure 3-7), and species with equilibrium levels above economic injury levels will be persistent pests (Figure 3-8) and will require constant pest control intervention.

While economic injury levels tell resource producers the maximum

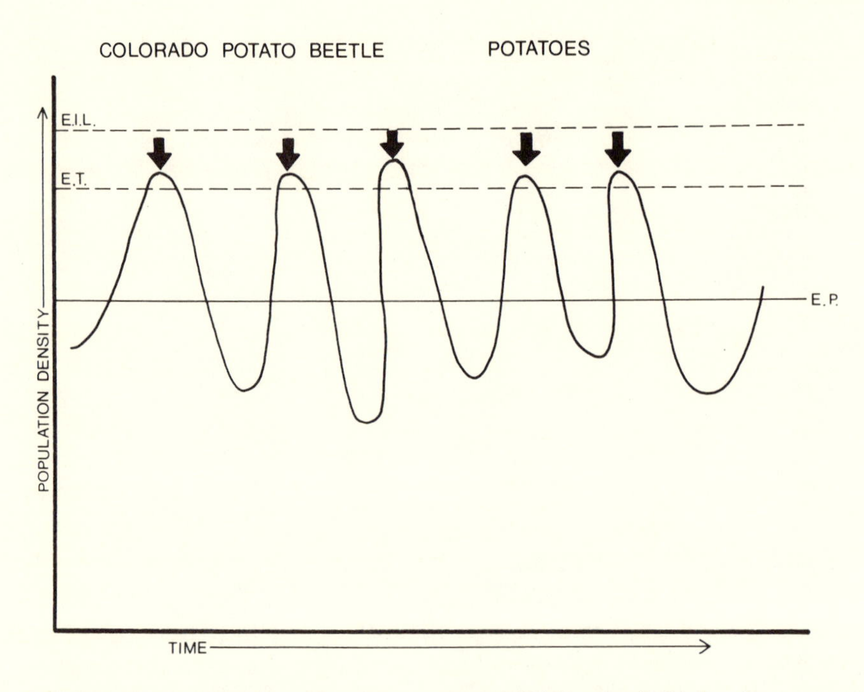

FIGURE 3-7. Equilibrium position close to economic injury level; this is a frequent pest requiring regular pest control action (after Luckmann and Metcalf, 1975).

numbers of a pest their crops can tolerate, they do not tell them when to do something about the problem. Control actions must be initiated before populations reach economically damaging levels so that significant loss can be avoided. Thus entomologists have devised a second population density level—known as the *economic threshold* or, more accurately, the *control action threshold*—which tells resource producers when they must take some action to prevent an impending pest outbreak.

Determining economic levels and control action thresholds is not a simple undertaking. Most pests are able to cause economic loss only during limited periods of time annually. One example is the tobacco budworm, *Heliothis virescens*. The budworm, as its name suggests, feeds on the tobacco leaf buds. Leaves so damaged at the bud become malformed and develop larger holes as they grow. However, by the middle of July, all leaves have popped out of their buds, and the budworm, at any density level, cannot inflict economic damage during the month remaining until harvest. Thus, economic levels have to be constantly adjusted in time according to the development or condition of the resource.

Likewise, accurate estimates of natural enemies must be made. Quite often a pest population boom to near-damaging levels will evoke a rapid response in its natural enemies and other elements of natural control that exert heavier pressures on increasing densities of the pest population, and a sudden crash of the pest population to a level below its equilibrium level will result. In this way the pest may be kept below economic levels in most places with no human action necessary. Knowledge of such relationships is vital to sensible pest management. We will discuss control action thresholds and their use in more detail in Chapter 7.

PLACING PESTS ON THE FOOD CHAIN

Proper identification of pests is the key to successful pest management. Far too often, pest management fails because an organism or group of organisms improperly assumed to be causing or threatening to cause economic damage is destroyed and the surrounding ecosystem is thrown out of balance—resulting in new pest problems. There are thousands of organisms in most sizable ecosystems, and as we discussed earlier, the vast majority of them are not pests and never become pests. How then

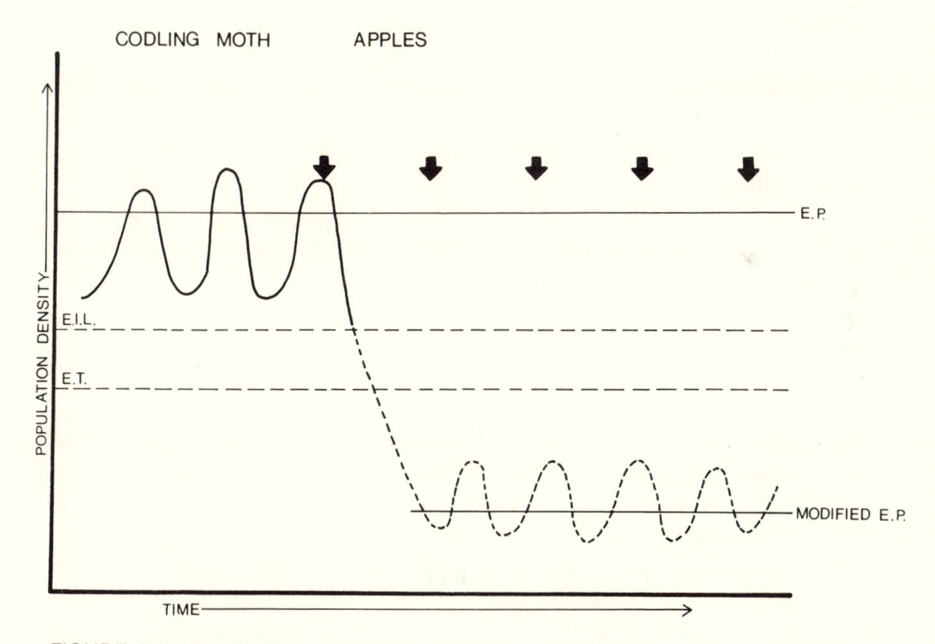

FIGURE 3-8. Equilibrium position above economic injury level; this is a constant pest requiring constant pest control action. Such constant pest control intervention can actually keep populations at a new *modified equilibrium level*. This modified equilibrium level should not be confused with a permanent lowering of the equilibrium level, as is illustrated in Figure 3-3. In this case, as soon as action is halted, pest populations will rise rapidly.

do we determine which organisms are causing our pest problems? A first step in determining this is to find out what each species eats. A simple food chain analysis can then show whether or not it could be a potential pest. Most pests fall into two categories along a resource-oriented food chain. They may (1) eat the resource—acting as *resource consumers* (e.g., leaf-eating caterpillars in a vegetable crop); or (2) have the same life-sustaining requisites as the resource (generally, food, water, light, and/or space) and compete with the resource for acquisition of the limited amounts of these essentials, thus functioning as *resource competitors* (e.g., brushy species in a newly replanted area of clear-cut forest). Figures 3-9 and 3-10 show examples of resource consumers and competitors in the production of plant and herbivore resources.

Organisms at higher trophic levels on the resource-oriented food chain—that is, those who eat the resource consumers or resource competitors— play a beneficial role in resource production. As natural enemies, these organisms often serve as density-dependent factors that lower the equilibrium levels of resource-consuming and competing organisms. Although their collective biomass will be smaller, the number of species feeding at these higher trophic levels is generally much greater than the number of herbivore species, especially in a monoculture situation. In other words, the food web gets more complex at higher trophic levels because most individual herbivores support several different species of carnivores.

This concept was clearly demonstrated in a study of the insects involved in the cabbage food web in Minnesota (Weires and Chiang, 1973). As we can see in Figure 3-11, the most common food habit among the insects in the cabbage food web is carnivory—the eating of other animals. The ratio of carnivores to herbivores in this food web is 3.4:1. Sugar solution feeders, the second largest category, are composed largely of adults of insects that are carnivores during their larval stages (including lacewings, ladybugs, and parasitic wasps) and also include a few adults of species, such as some moths and butterflies, that are herbivores as larvae. Weires and Chiang's study accounted for only the insect portion of the food web; the inclusion of the other invertebrate and vertebrate members of the food web, however, would probably not change the herbivore-to-carnivore ratio significantly.

With the knowledge that such a large proportion of organisms in the food web are carnivores and decomposers, we can understand the need to identify properly the feeding habits of organisms instead of merely assuming they are pests and automatically destroying them. We can also see why the feeding activities of many herbivores, with so many carnivores limiting their numbers, never constitute much of a pest problem. Just how important these natural enemies are is often not appreciated

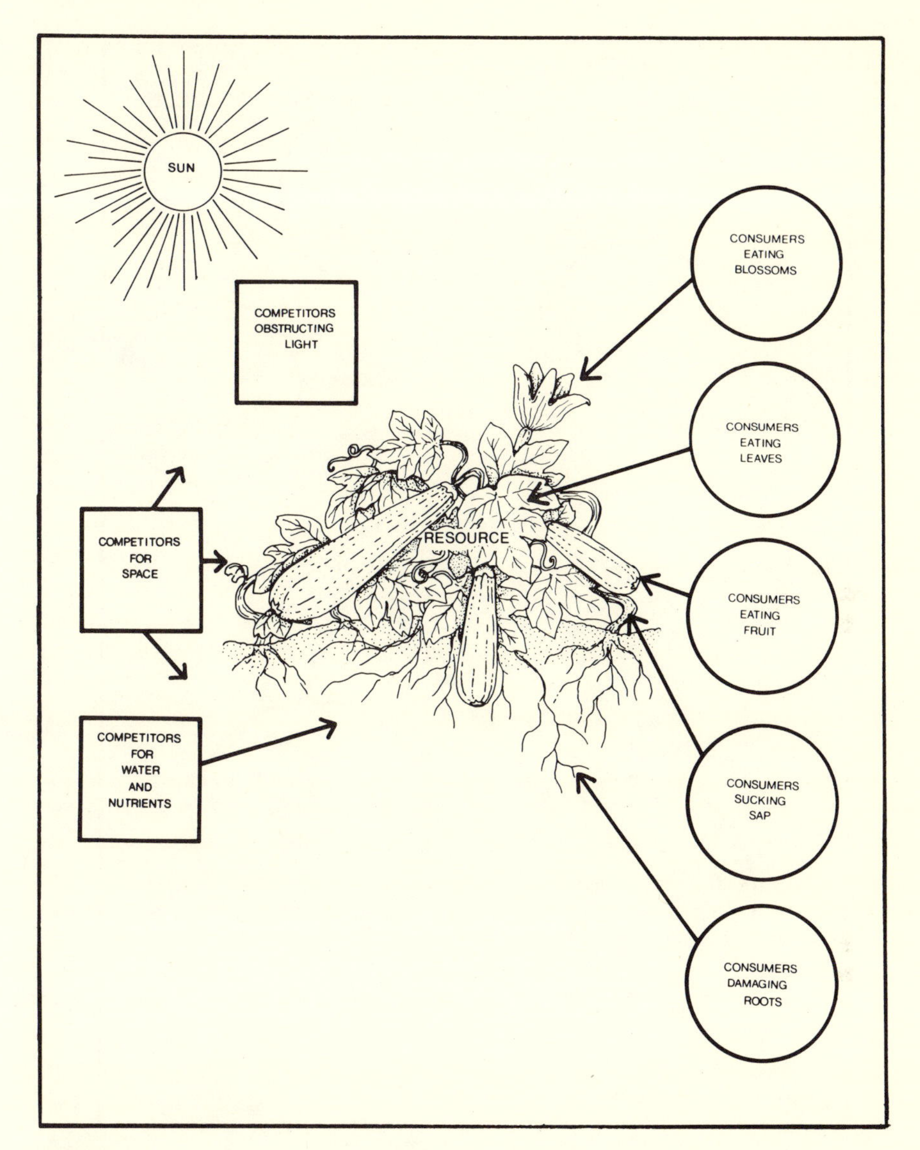

FIGURE 3-9. Possible roles of consumers and competitors in a plant resource ecosystem (zucchini).

until some action is taken (such as the ill-advised use of pesticides) that eliminates a beneficial species but not its prey, the potential pest. Huffaker and Kennett (1956) dramatically showed the value of one natural enemy in the strawberry agro-ecosystem by removing it with repeated insecticide (parathion) treatments. Figure 3-12 shows the phenomenal growth of the

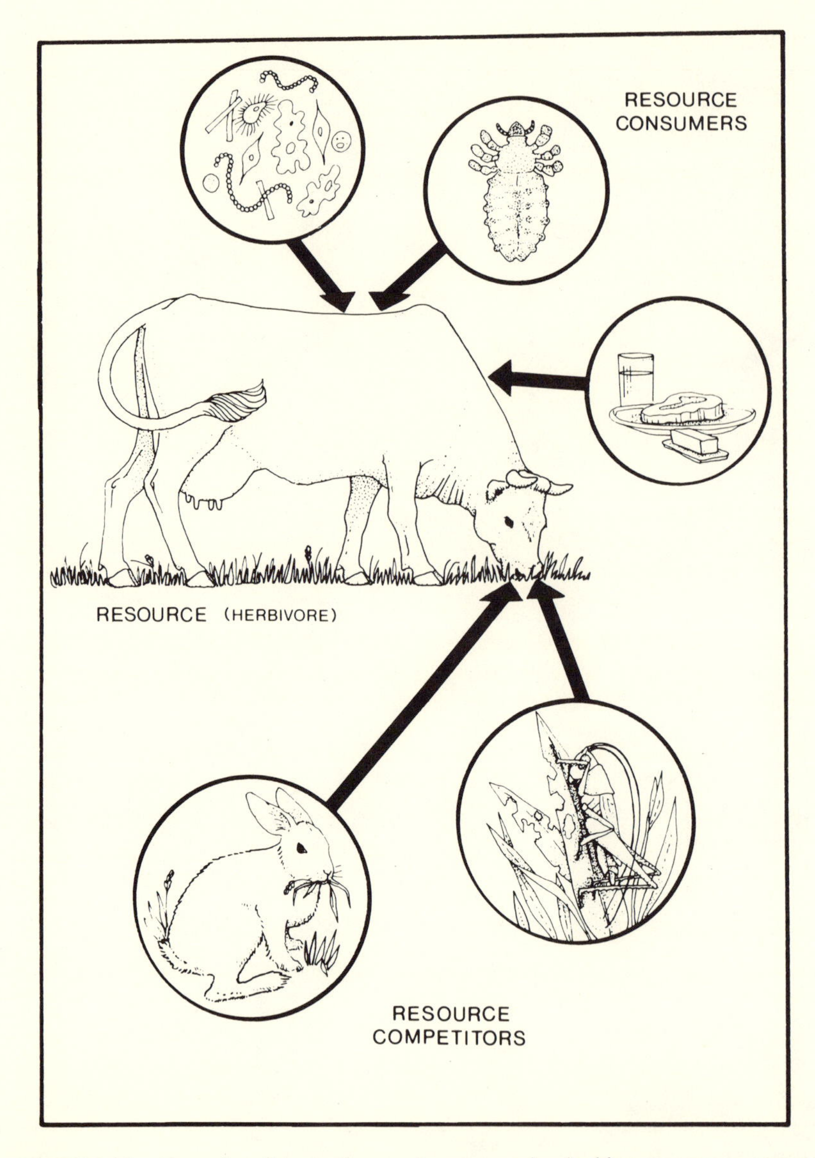

FIGURE 3-10. Examples of competitors and consumers in a herbivore resource ecosystem.

herbivorous mite (up to pest levels) in the parathion-sprayed, predator-free plots as opposed to predator-stocked, untreated ones, where this herbivorous mite did not become a pest.

Deciding precisely when resource competitors such as weeds become factors limiting resource yields is somewhat more complex than deriving economic levels for insect pests. However, weeds do become limiting factors in many situations. For instance, shading of crop seedlings by weeds can slow early growth considerably; obstruction of light by taller weeds can prevent the plant from realizing its optimum photosynthetic rate. Water is a limiting factor in many nonirrigated situations, and weeds can reduce substantially the amount available to crop plants. However, it is important to remember that many "weeds" contribute beneficial qualities to the managed ecosystem (e.g., food or shelters for natural enemies) that can, in some cases, outweigh their negative role as resource competitors.

Certain kinds of pests cannot be clearly defined as resource competitors or consumers. Disease-vectoring animals and "nuisance" plants and animals do not fit into a resource-oriented food chain analysis because

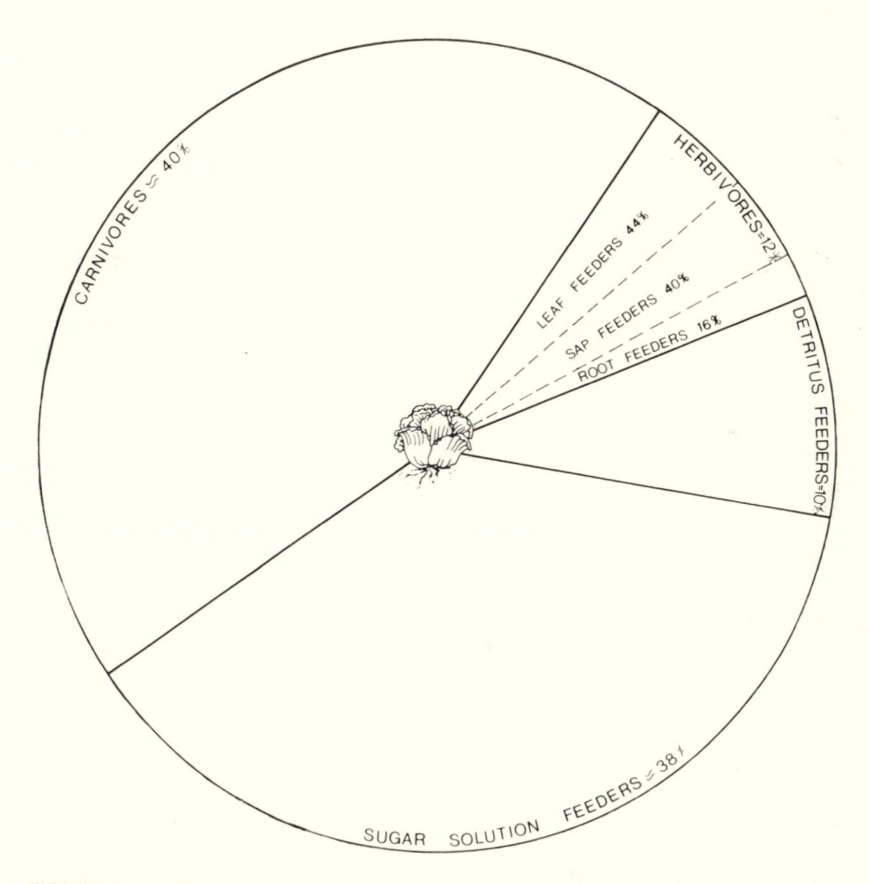

FIGURE 3-11. Food habits among the insect species in the cabbage food web (data from Weires and Chiang, 1973).

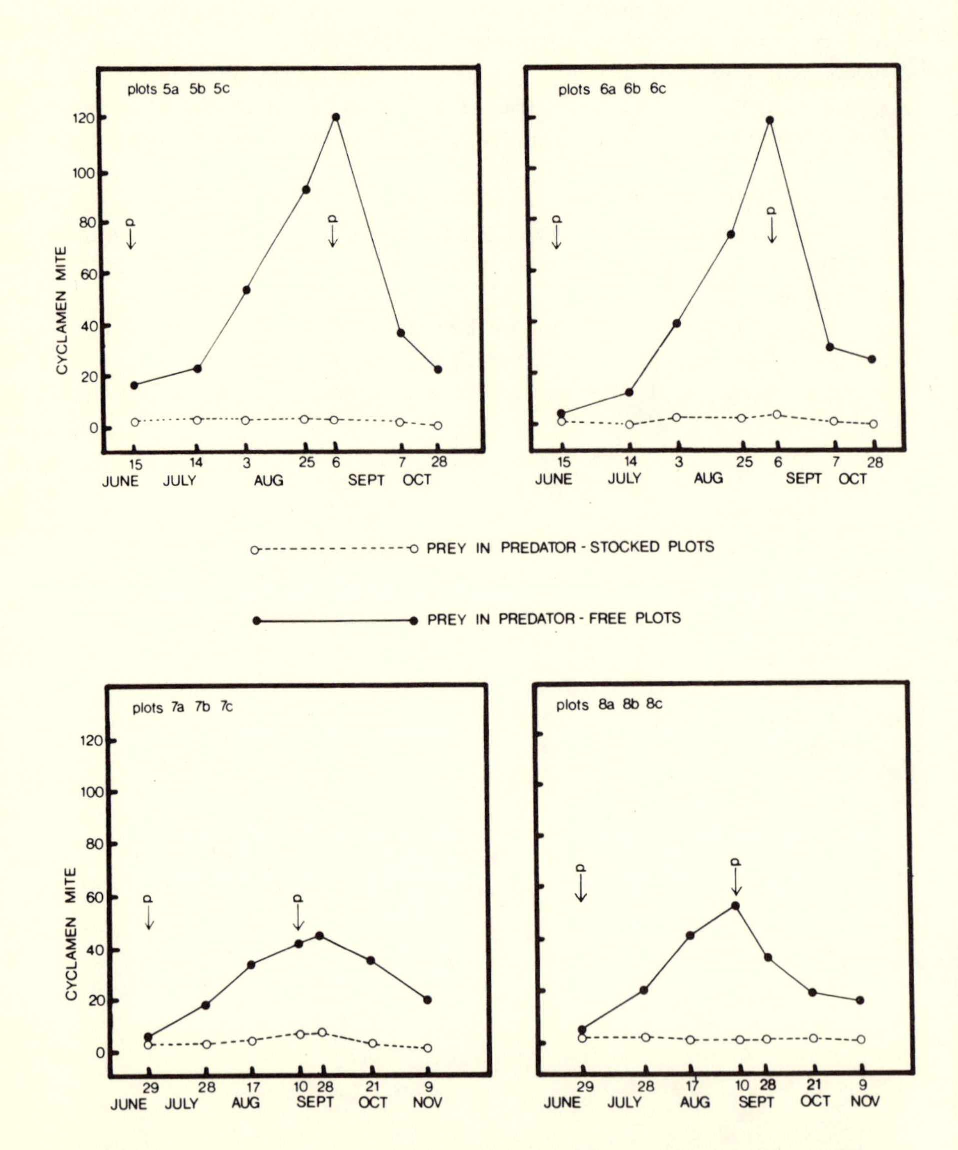

FIGURE 3-12. The importance of natural enemies in keeping a potential pest of strawberries (the cyclamen mite) from becoming a pest in different areas of California. Solid lines indicate populations of this pest where predators (phytoseiid mites) have been removed (by insecticide treatments). Dotted lines indicate adjacent plots where predators are present. The small *p*s with arrows indicate dates of insecticide (parathion) treatments in the predator-free plots (after Huffaker and Kennett, 1956).

they disrupt health and human comfort—hard-to-define resources. However, it is important to remember that these organisms, too, are part of similar ecosystem food chains with competitor, consumer, and higher trophic level (natural enemy) components.

KINDS OF PESTS

Organisms that cause a significant reduction in resource yield every season unless some pest management action is taken are known as *key pests*. These are the pests around which management systems are built. For instance, of the hundreds of insects occurring in one crop, only a very few species will be key pests. The rest either have no pest potential because of the nature of their biologies and feeding habits or will be kept from inflicting significant damage, at least most of the time, by natural control factors.

Occasional pests are organisms whose populations flare up to cause significant damage once in a while but generally do not cause intolerable loss. Their flare-ups are often due to disruptions in natural control, climatic irregularities, or mismanagement by man.

Potential pests comprise the vast majority of resource consumers and competitors. These organisms never cause enough damage to inflict significant loss of yield under prevailing management practices. However, because of their place on the food chain, they have the potential to become pests if management practices are changed. For instance, the cottony cushion scale is a potential pest in citrus in California; if management practices were changed (to cause large-scale killing of its natural enemy, the vedalia beetle), the cottony cushion scale would again become a serious pest.

Migrant pests cause occasional losses in a special way. These pests, such as locusts, some armyworms, and many bird species, are highly mobile and will infest the crop periodically for short periods of time, often inflicting severe damage.

Nonpests are those organisms in the ecosystem that have no potential for becoming injurious because of their place on the food chain relative to the resource. The effects of these organisms are often beneficial. For example, they may play an important role in the ecosystem by controlling pests, by recycling nutrients, by pollinating fruit or seed crops, or by providing alternative nutrition and shelter to beneficial organisms.

Certain organisms are considered detrimental in one situation and beneficial in another. A good example is the native blackberry bush in California. Although its tasty fruit feeds wildlife and an occasional hungry passerby, the blackberry bush is generally considered an undesirable weed, condemned for its nasty thorns and aggressive growing habit. Once established along a stream or in a vacant lot, the blackberry bush propagates rapidly through both seed and vegetative runners. In this manner, a single plant quickly proliferates into a large, impenetrable, thorny clump.

However, since 1961, grape growers in the Central Valley of California have learned that blackberry bushes have their beneficial aspects as well, especially in the control of an important insect pest—the grape leafhopper, *Erythroneura elegantula*. Long considered the key pest in many grape agro-ecosystems, this foliage-feeding leafhopper inflicts damage on leaves and fruit quality when present in large numbers. Insecticides have often failed to provide effective control of the leafhopper, or their use has aggravated other pest problems such as spider mites. Entomologists had known that a tiny natural enemy, the parasitic wasp *Anagrus epos*, which lays its eggs in the eggs of the grape leafhopper, kept the pest under control in some vineyards—but not in others. Nobody knew why.

The riddle was solved when it was realized that the wasp spent its winters parasitizing a different insect on a different plant host (Figure 3-13). Since the leaves fall off grapevines in the winter and the grape leafhopper retreats to the edge of the vineyard and becomes inactive, the nonhibernating parasitic wasp has no shelter, food, or means of survival in this environment. Nearby blackberry bushes, however, keep their leaves during winter and host their own leafhopper species, *Dikrella cruentata*, all year round. When entomologists checked the eggs of this blackberry leafhopper, they found considerable parasitism by *Anagrus*. Thus, the weedy blackberry patches were providing a winter home for this important natural enemy of the key grape pest.

Accordingly, it was growers with blackberry bushes in the vicinity of their vineyards who had the least grape leafhopper problems. *Anagrus* adults migrating back to the vineyards in the spring kept grape leafhoppers numbers at low levels from the beginning of the season. Since this discovery, a few growers have solved their major leafhopper problems with the planting of blackberry bush refuges in shady areas near their vineyards. Thus, the beneficial aspects of the blackberry bush in supporting natural enemies of a key grapevine consumer far outweigh its negative aspects as a grapevine competitor.

Similarly, the flowers and foliage of other "weedy" species provide life-supporting nectar, pollen, and shelter for other natural enemies of pest insects. The importance of these refuges has often become known after they have been removed. Then suddenly, large numbers of important

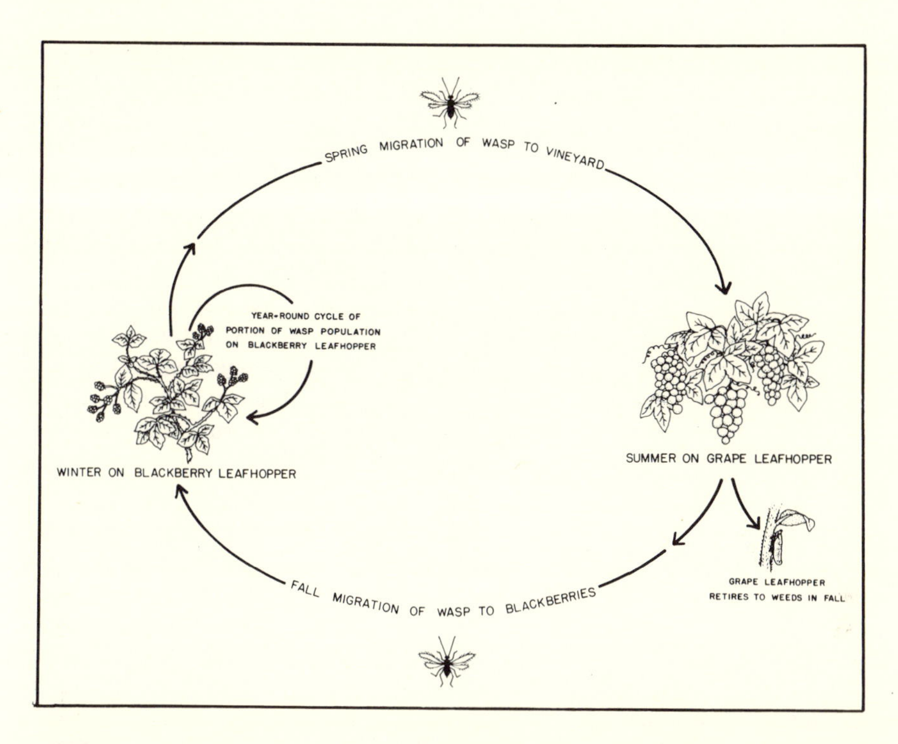

FIGURE 3-13. Life cycle of *Anagrus epos*, a parasite of the grape leafhopper. The parasite spends its winters in blackberry bushes parasitizing eggs of the blackberry leafhopper while the grape leafhooper remains inactive. In the spring, when the grape leafhopper is again active, part of the *Anagrus epos* population migrates back to the vineyard to parasitize the grape leafhopper's eggs. Some of the parasite's population remains in the blackberries year-round.

natural enemies die from starvation or exposure, and potential pests, released from their natural control, become actual problems for the first time. Such a sequence of events has been a common occurrence in orchards after the weeds beneath trees have been removed by plowing them under or by the application of a herbicide (see Chapter 7, Figure 7-15). In recognition of the role of these shelter and food sources in the ecosystem, the California Division of Highways and the University of California initiated a program testing shrubs that flower during the summer or all year round to provide nectar. Establishment of the appropriate flowering plants will allow maintenance of a higher level of parasite activity on roadside plantings throughout the year.

Resource consumers frequently play a positive role in the maintenance of natural enemies of a key pest, far outweighing the negative effects of their feeding. One such case is the apple rust mite, *Aculus schlectendali*,

appearance of even small numbers of a potential pest (and sometimes nonpest) species often results in an immediate desire in the home gardener to eradicate. Few people seem to understand how tolerant most ornamental plants are to moderate and even heavy damage (e.g., some trees can be completely defoliated two years in a row without dying), and even fewer would consider the simple alternative of replacing those plants that cannot tolerate such damage with other attractive ornamentals that can.

Entomophobia, the fear of insects, is epidemic in America, particularly among urban and suburban dwellers. This creates new pests yearly. When people cannot accept the presence of innocuous or even beneficial insects and other arthropods such as crane flies, spiders, and ground beetles, these species are classed as pests. The surest route to a solution to such "pest" problems is to introduce people to insects at an early age and to educate them about the fascinating and vital role these organisms play in maintaining the quality of our ecosystem.

CHAPTER 4

A History of Pest Control

EARLY PEST CONTROL—FROM PREHISTORIC TIMES TO THE RENAISSANCE

The organisms that disturbed prehistoric man's nomadic hunting–gathering life-style must have been few. As humans of this period grew no crops and had no permanent homes and few possessions, we can imagine that their pest problems would have been limited to those organisms, such as lice, fleas, flies, and mosquitoes, that caused people physical discomfort. Prehistoric control of these pests—picking, slapping, and squashing—could hardly be called a science.

It was only after the development of agriculture (approximately 10,000 years ago), the establishment of permanent settlements, and the introduction of a life-style that required the storage of greater or lesser quantities of food and other items that a concerted effort to control a large variety of organisms became necessary. Early pest control practices were often based on mysticism or superstition, such as an offering to a god or the performance of a ritual dance (Figure 4-1). But gradually, over the millennia and through the process of trial and error, a few useful methods became known—some are still successfully employed today.

Well before 2500 B.C., the Sumerians were using sulfur compounds to control insects and mites. By 1200 B.C., thousands of miles to the east in China, plant-derived insecticides (like present-day botanicals) had been developed for seed treatment and fumigation uses. The Chinese also used chalk and wood ash for prevention and control of both indoor and stored-product pests. Mercury and arsenic compounds were employed to control body lice and other pests. Interestingly, the beneficial role of natural

FIGURE 4-1. Enamel plate from ancient Ashur (now part of Iraq), representing an Assyrian noble in a locust prayer before the god Ashur, 650 B.C. (after Harpaz, 1973).

enemies and the value of adjusting crop planting times to avoid pest outbreaks had been recognized by the Chinese several centuries before Christ.

Similar techniques were in common usage among China's Greek and Roman contemporaries. In 950 B.C. Homer noted the value of burning for locust control. Herodotus (450 B.C.) mentions the use of mosquito nets and the practice of building high sleeping towers to avoid mosquitoes. A long-established use of fumigants by the Greeks was described by Aristotle in 350 B.C., and in 200 B.C., the Roman Cato reported the use of oil sprays, oil and bitumen sticky bands, oil and ash, and sulfur bitumen ointments for pest control. A pest-proof granary (similar to that in Figure 4-2) was designed by the Roman architect Marcus Pollio in 13 B.C.; it shows a clear understanding of the benefits of habitat modification in preventing pest problems. Additional protection from both mice and weevils could be obtained by treating the granary floor with a mixture of clay, chaff, and the fluid from oil presses.

However, not all pest control practices in the Roman Empire were as well founded biologically. Frustrated by seemingly insurmountable problems such as plagues of locusts or plant diseases, the people of early civilizations would recurrently turn to religion and superstition for help with their pest problems. For instance, a Roman agricultural text of 50 A.D. (*De re Rustica*) suggests the following for protection from caterpillars: "a woman ungirded and with flying hair must run barefoot around the garden, or a crayfish must be nailed up in different places in the garden." Also, the Romans traditionally performed certain rites in April to appease the goddess Robigo, who was identified with cereal rust diseases—the worst pest of the period (*Maxima segetum pestis*).

In China, the evolution of pest control technology continued during the first thousand years after Christ. It was favored by what was already

FIGURE 4-2. Galician "horreo" or granary. The design is unchanged from the time of the Celtic Invasion of Spain (ca. 500 B.C.). Made of granite slabs and wood, the horreo is fire- and vermin-proof. It rests on columns topped by circular stone rat guards. Rats are unable to climb upside-down around the stone guards and cannot get to the grain stored above. This practical protection from granary rat pests may have been the forerunner of the classical capital!

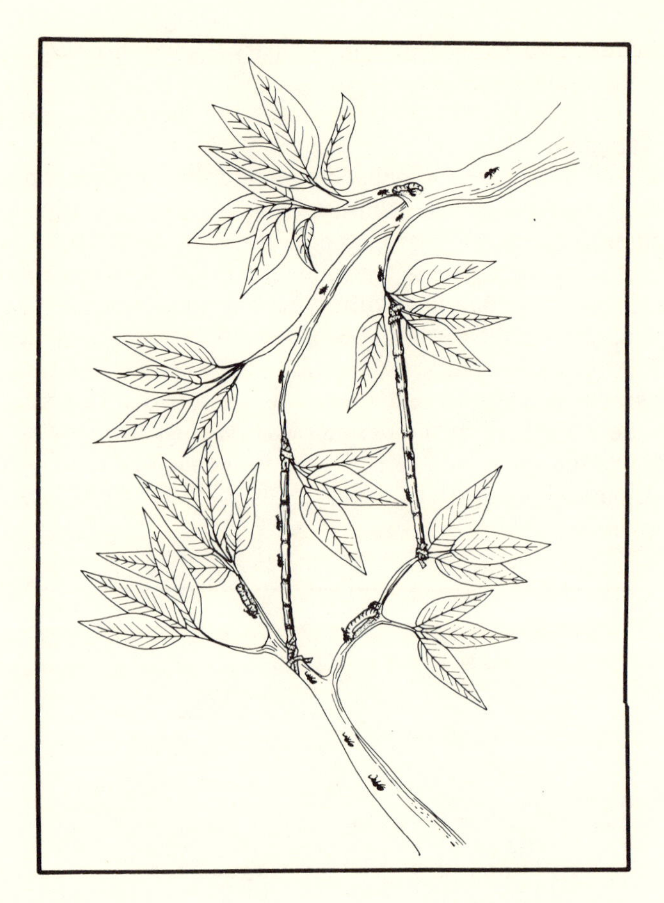

FIGURE 4-3. The earliest recorded use of natural enemies to control pest insects was the use of predatory ants in citrus orchards to control caterpillar and beetle pests in China. Nests were established in the orchards by Chinese growers, and bamboo bridges were placed between branches to facilitate the ants' movements from tree to tree.

a tradition of intense interest in and knowledge of insects (the cultivation of silkworms is reported to have been established in 4700 B.C.) and by a philosophical view of the world that early recognized food webs, feedback mechanisms, and other natural population controls. This understanding is well illustrated by the following passage from a Chinese text written in the third century A.D.:

> A factor which increases the abundance of a certain bird will indirectly benefit a population of aphids because of the thinning effect which it will have on the coccinellid (lady bug) beetles which eat the aphids but are themselves eaten by the bird. (Cited from Konishi and Ito, 1973.)

With such a basic appreciation for the functioning of ecosystems it is not surprising that the Chinese were responsible for the earliest application of biological control. It is recorded that by 300 A.D. the Chinese were establishing colonies of predatory ants in their citrus orchards to control caterpillars and large boring beetles. The ants' activities were facilitated not only by the strategic placement of nests in the orchards but also by the construction of bamboo runways that provided the ants with an easy transit route from one pest-infested tree to another (Figure 4-3).

Other pest control methods being employed in early China show a similarly remarkable sophistication in technique. Ko Hung, the great alchemist of the fourth century, recommended a root application of white arsenic when transplanting rice to protect against insect pests. Sulfur and copper were being used for lice control and pig oil was applied to protect sheep from parasites.

While China was continuing to advance its pest control approaches, European methods in the centuries after the fall of the Roman Empire relied increasingly on religious faith, superstition, and legalistic pronouncements, and less on biological knowledge. A few examples (from Dethier, 1976) illustrate:

571–630 A.D.	As protection against locusts, followers of Islam displayed prayers of Mohammed on poles in fields.
666 A.D.	St. Magnus, Abbot of Flussen, repulsed locusts and other pests with the staff of St. Columbia.
1476 A.D.	In Berne, Switzerland, cutworms were taken to court, pronounced guilty, excommunicated by the archbishop, and banished.

THE RENAISSANCE AND THE AGRICULTURAL REVOLUTION

In Europe, the Renaissance brought a rebirth of the search for scientific knowledge and an increased understanding of the organisms that became pests. The introduction of the compound microscope in the seventeenth century resulted in a burst of new information about man's various tiny competitors. Using the microscope, van Leeuwenhoek discovered bacteria in 1675. Other scientific advances in the seventeenth century included Redi's proof that insects do not arise spontaneously from decaying material but develop from eggs laid there, Valisneri's demonstration of the nature of insect parasitism, the discovery of human blood circulation by Harvey, and the recognition of the existence and function of numerous other organs in both humans and insects. In the first half of the eighteenth century, Linnaeus laid the foundation of true systematics with his development of the system of binomial nomenclature. Needless to say, a good system of nomenclature and identification was essential to the development of sound pest control.

During this period, approaches to pest control began to reflect greater biological understanding, although in many cases they were still limited in effectiveness (Figure 4-4). Reamur (1683–1756) discussed the significance of host–parasite relationships in pest outbreaks and suggested the use of entomophagous insects, specifically lacewings, to keep a greenhouse free of aphids. Later, Linnaeus suggested the use of ground beetles, ladybugs, lacewings, and parasites for the biological control of pests. He also advised the use of a predatory stink bug for control of bedbugs and the use of snails to reduce growth of moss on apple trees. Provision of nesting boxes for insectivorous birds in orchards and forests began to be a common practice in Germany in the early 1800s.

The late seventeenth and early eighteenth centuries also saw the rediscovery and/or introduction into Europe of various botanical insecticides: pyrethrum, derris, quassia, and tobacco leaf infusion—all effective insecticides. The dangers of the use of other toxic poisons became known in the 1700s as well. In 1754, Aucante in France observed arsenic poisoning among agricultural field workers, and in 1786 France prohibited the use of arsenic and mercury steeps for seed treatment.

The period 1750–1880 in Europe was a time of agricultural revolution. Farming, for the first time, began to become more of a commercial than subsistance enterprise: average yield changed from four seeds produced per one seed planted to ten seeds per one planted. This increase in yield was due largely to changes in land distribution and agricultural practices. These changes included a reorganization of land holdings that eliminated the landlord–serf relationship, the expansion of planting acreages, and the introduction of new farming techniques such as sophisticated manuring practices and good rotation systems involving nitrogen-fixing fodder crops. During the mid-1700s farmers began to grow crops in rows, thus permitting weed removal with the horse-pulled hoe.

The greatest single cause of large-scale crop disaster then, as it is now, was not pests but weather. Weather-induced damage may be "direct" (e.g., drought, flood, early freeze, tornado) or "indirect" (e.g., wheat diseases such as rusts and black scab, which are favored by high humidity) by providing an environment conducive to the development of epidemics of plant diseases and other pests.

As the period of "agricultural revolution" was peaking in the mid- to late nineteenth century, European countries and their colonies experienced some of the worst agricultural disasters ever recorded: the potato blight in the late 1840s in Ireland, England, and Belgium; the outbreak of powdery mildew in the 1850s in the grape-growing areas of Europe;

FIGURE 4-4. An eighteenth-century flea trap to be worn around the neck. Fleas entered the outer perforations (bottom left) and were caught on a sticky tube inside (top left). No record of the effectiveness of this trap remains. However, we do know that fleas were a constant harassment to people of all classes during this period in Europe.

the epidemic of the fungus leaf spot disease of coffee, which caused Ceylon to switch from coffee production to tea cultivation; and the invasion of Europe by an American insect, the grape phylloxera, which nearly put an end to the wine industry in France (1848–1878). Undoubtedly, the unprecedented severity of these outbreaks was due, at least in part, to the new, larger-scaled, commercially oriented farming practices and to newly arrived pests or strains of pests brought in by increased international travel.

Predictably, during this period there was a sudden surge of interest in perfecting pest control techniques. The first books and papers devoted entirely to pest control began to appear in the early nineteenth century. The first textbook on plant pathology was published in 1858 (by Kuhn) and listed climatic and soil conditions, insects, parasitic higher plants and microorganisms as causes of plant diseases. However, at this time (unlike in insect pest situations) there were few useful controls for plant diseases. Thaddeus William Harris's *Treatise on Some of the Insects Injurious to Vegetation*, published in 1841, was the first textbook in America on the control of insects, and it continued to serve as the prime source book for such information up until the 1870s. Examples of some of his pest control recommendations are listed in Table 4-1.

Although Harris's remedies are often labor-intensive and time-consuming relative to today's management programs, they were at least partially effective and reflected a broad knowledge of pest biologies and pest–host interactions. Suggested remedies ranged from hand-picking and shaking (Figure 4-5), encouraging natural enemies, employing various cultural practices (e.g., adjusting planting time to disadvantage the pest, enhancing plant growth and vigor with manure fertilization, sanitary practices such as burning after harvest, selection of pest resistant varieties), constructing physical barriers to pests (e.g., tree-banding with sticky substances), to the use of toxic and noxious substances (including whitewash and glue, tobacco, walnut, hops and other plant infusions, sulfur, soapsuds, whale oil, resin and fish oil, and lime and turpentine).

Two pests nearly devastated the European wine industry in the latter half of the nineteenth century—an epidemic of powdery mildew and the introduction of an insect from America, the grape phylloxera. The solution of these problems marked a turning point in pest control. These were probably the first major pest outbreaks in which human-directed efforts played the primary role in the control and containment of the pest. They also spurred the evolution of methods and pest control that were to dominate the scene for another fifty years.

The attack on the grape phylloxera problem was multifaceted. A phylloxera-eating mite was imported from America and established in 1873 but failed to become an effective control for the pest. Attempts at chemical control proved uneconomical and ineffective. A new approach, the use of an insect pathogen, was suggested by Louis Pasteur in 1874 but was never actually attempted. Pasteur's main research project at the time was the control of pebrine, a disease of the silkworm that was plaguing the silk industry in France in the last quarter of the nineteenth century. Pasteur wondered why such virulent diseases couldn't be put to some good use.

The breakthroughs that actually eliminated the phylloxera as a serious pest in Europe, however, were the successful utilization of host plant resistance and the evolution of the technique of grafting. A variety of American grape that was resistant to the phylloxera was discovered around 1870. This discovery was followed by the development of the

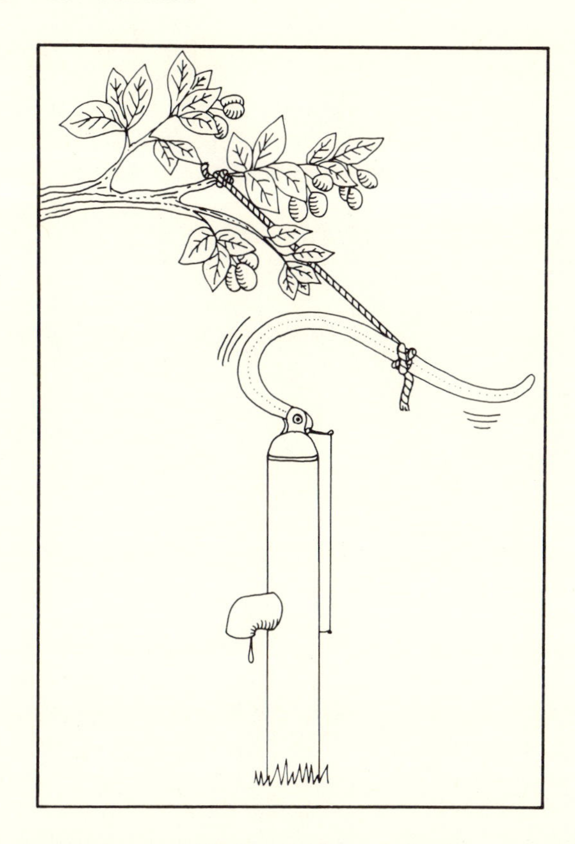

FIGURE 4-5. An ingenious method of control for curculios (a type of weevil with boring larvae) injuring plum and apricot trees, devised by Colonel T. Forest of Germantown, Pennsylvania, in the early 1800s. "Having a fine plum tree near his pump [he] tied a rope from the tree to his pump handle, so that the tree was greatly agitated every time there was occasion to pump water. The consequence was that the fruit on his tree was preserved in the greatest perfection" (from Dethier, 1976).

grafting technique that allowed the rootstocks from these resistant grape varieties to be joined on the popular European varieties. The resulting grafted plants did not suffer serious damage from the phylloxera.

The solution to the powdery mildew fungus problem came about by accident. A farmer, in an attempt to stop the pilfering of his grapes by passersby, applied a poisonous-looking mixture of copper and lime to his roadside plants. On later examination, he discovered that these roadside plants had escaped infection by the fungus. This "accident" resulted in the development of two fungicides that were to dominate plant pathogen control for many years to come. These were Bordeaux mixture (hydrated lime plus copper sulfate), still the most widely used fungicide in the world,

TABLE 4-1
Suggested Insect Controls from T. W. Harris's *Treatise on Some of the Insects Injurious to Vegetation* (1841)

Coleopterous borers on trees	Protect the woodpecker
Pine weevil	Cut off shoots in August and burn them; stick cut branches in the ground in the egg-laying season to trap eggs
Plum curculio	Jarring; gathering fallen fruit, spraying fruit with whitewash glue
Pear-tree scolytus	Pruning
Apple-borers	Clean culture; put camphor in plugged holes
Flea beetles	Sprinkling with tobacco and red pepper; watering with Glauber's salt and water; tobacco water; infusions of elder leaves, walnut leaves, hops, ground plaster of Paris, charcoal dust, powdered soot, sulfur, and Scoth snuff; torches; covering with millinet on frames
Cockroaches	Poison baits
Mole crickets	Poison baits and pigs
Squash bugs	Early hand picking and forcing the growth of plants by manuring
Vine leafhopper	Fumigation with tobacco under a movable tent, syringing with whale-oil soap and water
Aphids	Solutions of soap or a mixture of soapsuds and tobacco water used warm; also hot water; one-half ounce of carbonate of ammonia to one quart of water; lime; fumigating with sulfur or tobacco
Scale insects	Two parts of soft soap in eight parts of water mixed with lime to make a whitewash; two pounds of potash to seven quarts of water; one quart salt to two gallons of water
Peach-tree borer	Remove the earth around the base of the tree, crush the cocoons and borer, cover the wounded parts with moist clay, surround the trunk with a strip of sheathing paper extending two inches below the level of the soil, and place a fresh mortar around the root to confine the paper (do this in the spring or in June)
Hairy caterpillars (woolly bear caterpillar and allies)	Pay children to collect them by the quart
Salt marsh caterpillar	Mow the marshes early in July and, if possible, for several years in succession; burn over marshes in March
Cutworms	Soaking of the grain before planting in copperas water; rolling the seed in lime or ashes; mixing salt with the manure; fall plowing of sward lands intended for

TABLE 4-1 (*Continued*)

Cutworms (*continued*)	wheat or corn the following year; manuring soil with sea mud; protecting cabbage plants by wrapping a walnut or hickory leaf (or paper) around the stem
Cankerworms	Tree banding with clay mortar, strips of old canvas or strong tarred paper, a collar of boards smeared with tar, collars of tin plate, a belt of cotton wool, or troughs of tin or lead filled with cheap fish oil; melted Indian rubber; dusting leaves when wet with dew with air-slaked lime; one pound of whale-oil soap to seven gallons of water used as a sprinkle with a garden engine; jarring the trees; use of pigs to destroy pupae under the ground
Codling moth	Gather windfalls; wind cloth around the tree or hang in the crotches to attract larvae ready to spin; scrape off the loose and rugged bark; drive away the moths at egg-laying time by smoke of weeds burned under the tree
Clothes moths	Expose garments, furs, or feathers to the air and to the heat of the sun for several hours, then brush, beat, and shake before packing away. Brush over walls and shelves of closets with spirits of turpentine. Put powdered black pepper under the edges of carpets. Place sheets of paper sprinkled with spirits of turpentine, camphor, or coarse powder, leaves of tobacco, or shavings of Russian leather among clothes when put away in summer. Put small articles into brown paper bags securely closed, also put in a few tobacco leaves or bits of camphor. Use chests of camphor wood, red cedar, or Spanish cedar. Cloth linings of carriages: wash or sponge on both sides with a solution of corrosive sublimate of mercury in alcohol strong enough not to leave a white stain on a black feather. Fumigate with tobacco smoke or sulfur. Expose to steam for fifteen minutes. Place the infested garment in an oven heated to 150°F.
Angoumois grain moth	Heat for twelve hours at 168°F; early threshing and winnowing of wheat
Jointworms	Burn stubble, also straw and refuse; manuring and thorough cultivation, promoting rapid and vigorous growth of the plant
Hessian fly	Selection of varieties; burning the stubble
Horseflies	Protect animals by washing their backs with a strong decoction of walnut leaves

and Paris Green (copper acetoarsenite). Both were later found to have important insecticidal attributes as well, and Paris Green became one of the most commonly employed insecticides in the late nineteenth century.

Materials used for chemical control of pests did not change much in the fifty years after 1880. The active ingredients in most of these materials were compounds of arsenic, antimony, selenium, sulfur, thallium, zinc, copper, or plant-derived alkaloids. Hydrogen cyanide gas was also in use for fumigation purposes and various oils were used in the control of pests. Over the next decades these products were refined and made more useful by development of better application devices and techniques, by better timing of applications, and by the addition of inert but useful agents to facilitate surface adhesion and even distribution of materials.

Chemical control of weeds found its first application in 1896 when iron sulfate was found to kill broad-leaved weeds but not cereal crops. Over the next ten years many other simple inorganic compounds—e.g., sodium nitrate, ammonium sulfate, and sulfuric acid—were put into very limited use as herbicides. However, at that time labor was so inexpensive that few farmers were interested in chemical methods of weed control. Most depended on a combination of clean cultivation, tillage, crop rotation with weed-competitive crops, and hand weeding to keep their weeds pests under control.

It was also in the late nineteenth century that the importation and establishment of natural enemies for biological control was shown to be one of the most effective means of combatting insect (and later, weed) pests. The first major success of this technique was in the control of the cottony cushion scale in California. The cottony cushion scale was accidentally introduced into California in the late 1860s; by the 1880s it had spread throughout the citrus growing areas in California and was threatening to wipe out that industry. The native home of the pest was determined to be Australia. With this assumption, the United States government sent an entomologist, Albert Koebele, to Australia to send back natural enemies of the scale to be established in California. Of the natural enemies he found in Australia, Koebele sent back two: a parasitic fly, *Cryptochaetum iceryae*, and the vedalia beetle, *Rodolia cardinalis*. The vedalia beetle (Figure 4-6) turned out to be an exceptionally fast and effective control. One hundred and forty of these predators were carefully shipped back to California and turned loose on a screened-in cottony cushion scale-infested orange tree. Within a year and a half, the descendants of these 140 beetles had checked the cottony cushion scale over the citrus-growing areas of the state. Control by the vedalia beetle has been so successful that since its establishment in 1890, the cottony cushion scale has never (with one exception—see page 76) risen to pest status

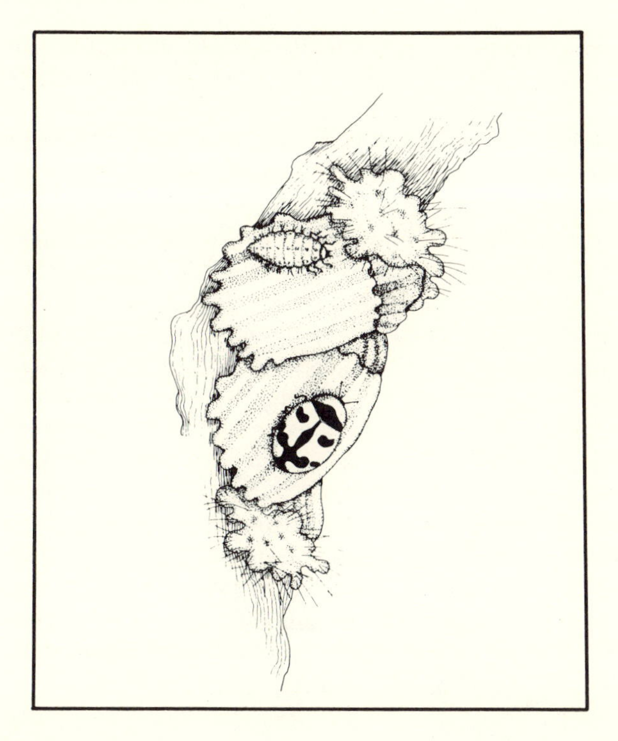

FIGURE 4-6. The vedalia beetle and larva feeding on the cottony cushion scale.

again. The introduced parasitic fly also became an important control factor in coastal areas of Southern California.

The 1890s saw an incredible succession of breakthroughs in medical entomology. It was during this period that arthropods were proven to be carriers (or "vectors") of disease organisms. The first proof was Smith and Kilborne's work in 1893 with Texas cattle fever, a protozoan disease of cattle, showing it to be tick-borne. The rest fell like dominoes: Bruce, in 1896, showed the African sleeping sickness pathogen to be carried by tsetse flies; rat fleas were shown to harbor the plague or "black death" bacterium and mosquitoes were identified as vectors of the malarial protozoa in 1897; the role of flies in the mechanical transmission of typhoid fever was proven in 1898; and in 1900 mosquitoes were positively identified as carriers of the yellow fever virus.

Accordingly, it became apparent for the first time that many serious diseases could be contained through the control of their arthropod vectors. The control of these disease-transmitting animals grew into a whole new area of pest management. The building of the Panama Canal (completed in 1915) represented the first large-scale success in controlling a medically important insect vector. The failure of the French in their attempt to build the canal in the last quarter of the nineteenth century can be at least partially credited to their inability to control malaria and yellow fever—due primarily to their ignorance of the role of mosquitoes as vectors. Mosquito control in the early 1900s focused on destruction of breeding sites by draining, filling, impounding, and periodic flushing and occasionally involved the use of a larvicide such as kerosene.

By the turn of the nineteenth century, five major approaches to pest control were well established and in common use: (1) biological control, (2) mechanical and physical control, (3) cultural control, (4) chemical control, and (5) use of resistant varieties. Pest control practice today still relies almost entirely on the utilization of methods in these five categories. Advances in the last 75 years have resulted largely from modifications of materials and practices in these areas, the introduction of new materials and techniques, and the successful employment of practices in two or more of these areas simultaneously and consciously for improved and longer-lasting pest prevention and control. A sixth approach, legal control, through the use of inspections and quarantines to prevent the entry and spread of pest-infested materials, was firmly established in the United States in 1912 by the Plant Quarantine Act of that year.

THE EARLY TWENTIETH CENTURY

By the early 1900s the number of people actively employed as economic entomologists, plant pathology experts, and other pest control specialists was substantial. Textbooks from that period show that these sciences were well developed. A close look at an entomology text of the period, E. Dwight Sanderson's *Insect Pests of Farm, Garden and Orchard*, published in 1915, reveals considerable progress in insect control thinking from Harris's haphazard approach of 1841. Sanderson shows us an approach that is systematic, well formulated, biologically based, and clearly thought out in a stepwise fashion in case after case; it is an approach to pest control that could be instructional and appropriate for students even today. Texts of this period stressed the importance of correct identification of pests and the need for a solid understanding of pest biology, especially in the timing of application of control measures.

Sanderson's 1915 text considered proper farm methods as a key to good pest control. These methods included crop rotation, arrangement of planting times to avoid pest outbreaks, and the destruction of weeds or "volunteer" crop plants, which might maintain pest populations during period of crop absence. Sanderson also pointed out the importance of

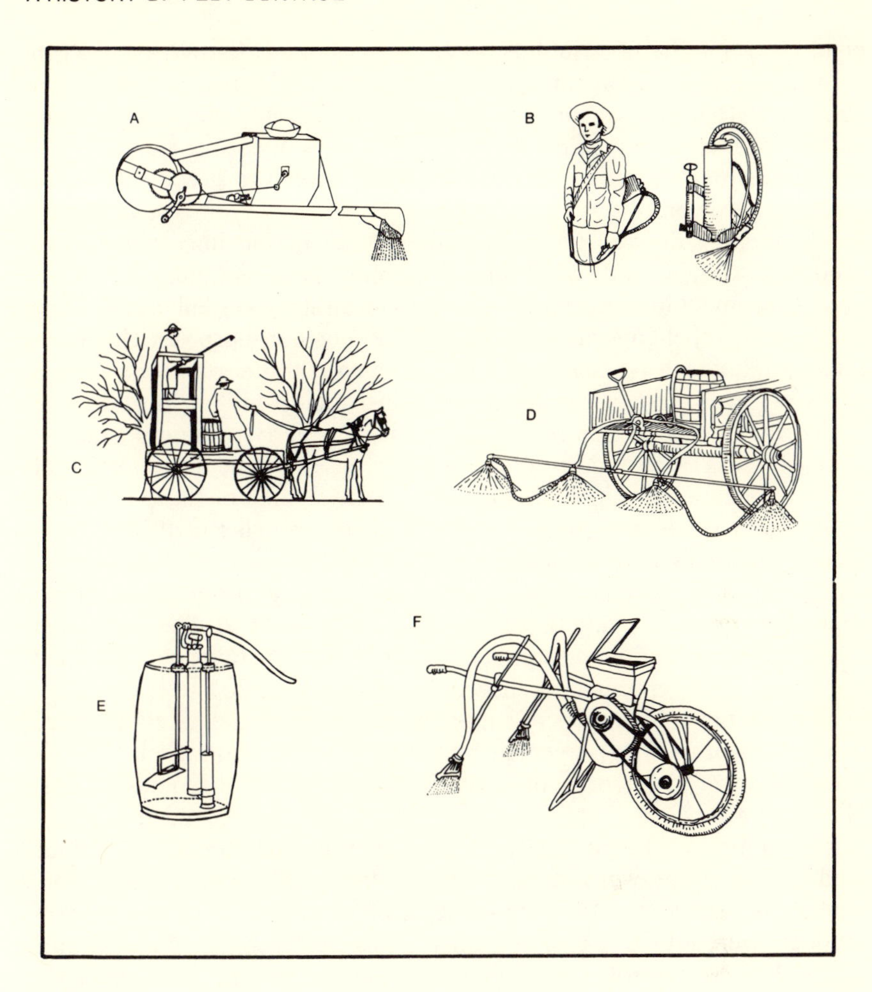

FIGURE 4-7. Early pesticide application equipment, ca. 1915. (A) A powder-gun for applying insecticides in dust form. (B) Compressed-air sprayer, individually held. (C) Spraying orchard trees from a rough tower bolted to a one-horse wagon. (D) Row-spraying attachment for use with barrel pump, adjustable for various widths of rows. (E) Barrel pump. (F) Wheelbarrow applicator for dusts.

proper fertilization and soil preparation in pest control, noting that a healthy crop can better withstand pest injury. He advocated the practice of "clean farming" by destroying fall stubble and refuse in which such pests as the corn stalk borer, the cotton boll weevil, and the chinch bug might overwinter. He also suggested the use of "trap crops" to attract pests away from the economically important crops—e.g., the use of corn to lure the egg-laying female *Heliothis zea* moth away from the cotton plants. Corn is the favorite host of *H. zea* (variously known as the corn earworm, cotton bollworm, and tomato fruitworm) but it is usually not available late in the season, so the moths are forced to seek out cotton. Sanderson recommended planting just a few rows of corn in a cotton field (to be destroyed after egg-laying had taken place) to greatly reduce a bollworm problem.

Sanderson divided the insecticides of the period into four classes according to their mode of action: (1) stomach poisons (killing via ingestion), (2) contact insecticides (clogging up respiratory system or corroding cuticle), (3) repellents, and (4) gases for fumigation purposes. Lead arsenate, a stomach poison highly toxic to man and other animals as well as insects, was the most commonly used insecticide until the introduction of fluorine compounds in the 1920s. During the early twentieth century, the usefulness of pesticides was greatly increased by the development of better application equipment. Figure 4-7 shows some of these early designs. The airplane was first used for pesticide spraying in 1921 in Ohio against the catalpa sphinx moth.

Physical and mechanical control devices were important aids in insect control in the early twentieth century (Figures 4-8 and 4-9). These included the screening of houses, mosquito nets, and the use of sticky bands around tree trunks to keep climbing insects from getting up to the leaves. The invasion of fields by hordes of land-migrating insects such as chinch bugs and army worms was prevented by constructing barriers of oils, dusty-sided furrows, low fences of sheet metal, and other ingenious means. Mechanical devices such as hopper dozers were employed to catch a variety of insects. Flypaper, fly traps, moth traps, light traps, and various kinds of bait traps were in common use during this time. Physical manipulation such as the flooding of fields or heating or cooling of stored products were effective pest controls. It was well recognized at this time that a physical control—the draining of swamps, marshes, and other accumulations of standing water—was the most effective method of destroying mosquitoes and horse flies.

Advances in plant pathogen control during the first four decades of the twentieth century occurred in several areas but were dominated by the establishment of plant breeding for resistance as an active area of research. Early plant breeding successes included the development of resistance to rusts in cereals and to *Fusarium* wilts of cotton, watermelon, and cowpea. Crop rotation and crop refuse destruction were recognized as effective in controlling many plant pathogens. Bordeaux mixture continued to be the leading fungicide, although several organic compounds such as the organomercuries, salycylanilide, and the dithiocarbomates had been introduced before 1940.

Progress in weed control proceeded on several fronts. The feasibility

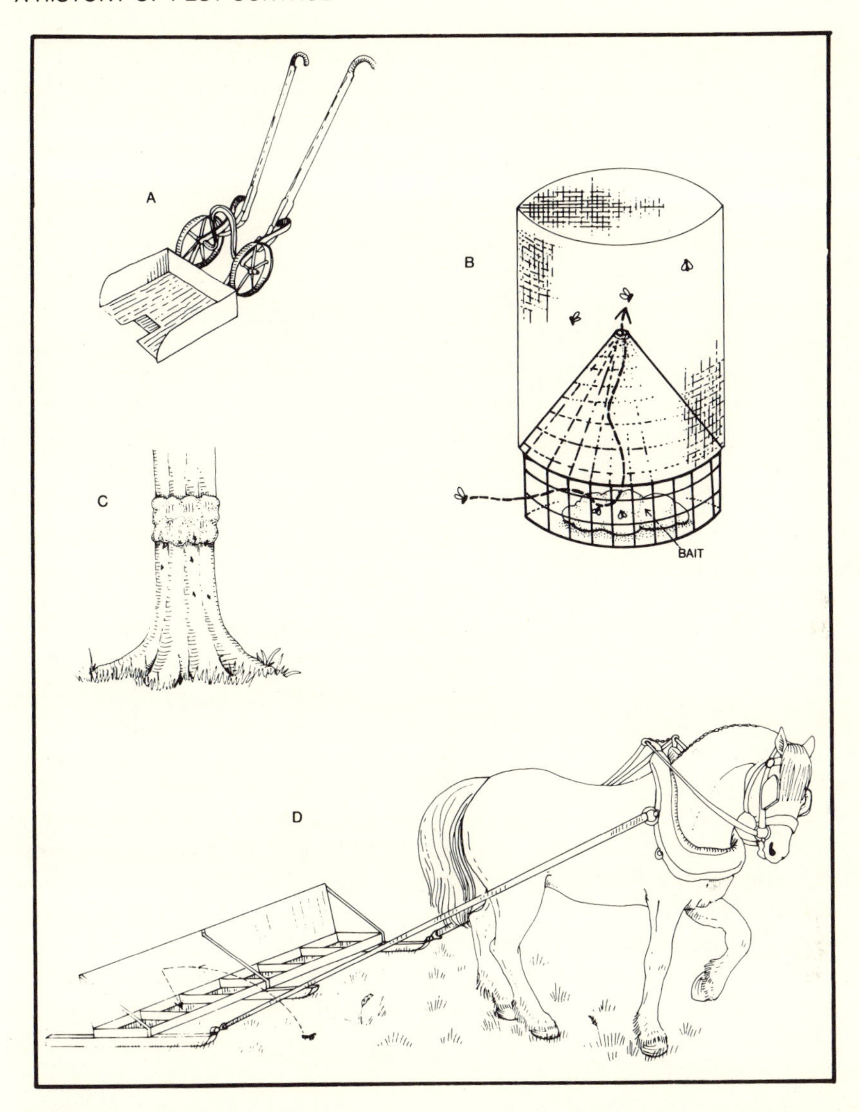

FIGURE 4-8. Physical and mechanical control devices of the early twentieth century. (A) A hopperette, designed for catching leafhoppers. Immediately after weedy areas, grass, or forage crops are cut, a hopperette can be pushed through the infested area; thousands of leafhoppers will fly into the machine and adhere to the sticky substance on its sides and bottom. (B) A fly trap. Flies are attracted to bait in the bottom of the trap, then fly up into the cone and cannot get out. With an attractive bait and a correctly sized trap, buckets of flies can be caught in a short period of time. (C) Sticky band or "tangle foot" around the trunk of a tree. Insects migrating up into the leafy portions of the tree get stuck in the band. (D) A hopperdozer. This is similar to the hopperette described in (A), but it is larger and designed particularly for catching grasshoppers. Oil or kerosene is placed in the trough of the hopperdozer to kill the pests once caught.

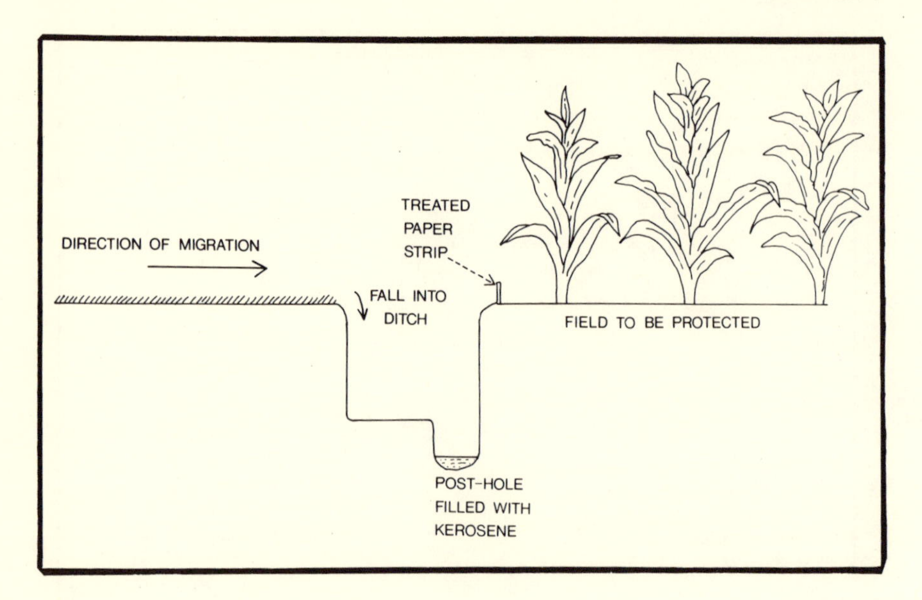

FIGURE 4-9. Paper strip barrier and post hole trap for protection of crops from migrating insects (e.g., chinch bug nymphs, migrating larvae of army worms). Migrating insects fall into ditch and are killed in kerosene. Tarred paper, saturated with creosote, repellent to chinch, is an additional deterrent preventing these insects from being blown across the barrier.

of controlling weeds by biological means was dramatically illustrated in Australia in 1926 with the successful introduction of the *Cactoblastis* moth and other cactus-feeding insects for the control of prickly pear cactus. Within ten years, over 60 million acres had been cleared of this nasty pest. The establishment of standards for weed-free seed and the development of better farm equipment and cultivation methods were important components of weed control during the first decades of the twentieth century.

AFTER WORLD WAR II—THE REVOLUTIONIZING OF PEST CONTROL BY DDT AND OTHER SYNTHETIC ORGANIC PESTICIDES

The science of pest control progressed steadily during the first 40 years of the twentieth century; but it was the pressures presented by World War II that caused the greatest revolution in twentieth century pest control—the development of the synthetic organic pesticides.

World War I had been fought primarily in Europe, and the pest problems that plagued the fighting troops there were the usual uncomfortable but rarely serious problems—lice, fleas, bedbugs—brought on by the inevitable crowded and often unsanitary wartime conditions. However, much of World War II took place in the tropics, and the insect-vectored diseases in these areas—malaria, typhus, sleeping sickness, dengue, relapsing fever—had the potential of becoming truly devastating to the entire war effort. Both sides realized this immediately. Research on more effective insecticides became a top priority.

In the United States hundreds of chemicals from manufacturers around the world were put through a screening process for insecticidal activity. One of these routinely tested materials, dichloro-diphenyl-trichloroethane (DDT), manufactured by the Geigy Chemical Company of Switzerland and developed by a Swiss chemist, Paul Mueller, was just what the researchers had been looking for—a substance toxic (even in minute quantities) to virtually every test insect! The wartime benefits conferred by DDT and similar insecticides, through their diminution of various diseases, should not be underestimated.

While the Western Allies were developing the chlorinated hydrocarbons, the Germans had come up with another equally toxic group of insecticidal compounds—the organophosphates (including HETP, parathion, and schradan). A third group of synthetic organic insecticides, the carbamates, was also discovered in the 1940s by Swiss workers; but these materials did not come into popular use until the late 1950s, with the development and marketing of carbaryl in the United States. The first use of these new insecticides was, of course, for control of insects that carried human disease. But after the war they found a ready market in peacetime agricultural enterprise. Their success was immediate. They were cheap, effective in small quantities, easy to apply, and widely toxic. They seemed to be truly "miracle" insecticides.

The pesticide industry boomed. In the early 1900s, pesticides were usually mixed up in the back shed by the farmer himself, according to directions in the latest farm journal. As pesticides came into more widespread use and became regulated, small industries specializing in pesticides sprang up. But with the coming of the synthetic organic pesticides and large numbers of petroleum-derived products, some of the world's largest corporations became involved in the development, manufacture, and marketing of pesticide products. The introduction of selective herbicides (such as 2,4-D) in the 1940s was soon followed by the development of low-volume sprayers and other field application equipment and technology. Thus, the application of pesticides, a practice confined largely to orchard and high cash crops, became a common procedure in just about every agricultural crop and, subsequently, in urban and recreational areas as well.

The effect of the new pesticides on the attitude of those who controlled pest organisms was revolutionary. Where farmers had formerly talked of "controlling" pests, expecting to have to tolerate certain levels of the noxious species, they now talked of "eradicating" pests. People envisioned the extermination of entire species of pest insects, plant pathogenic organisms, and weeds, and expected 100% kill from their pest control actions. The new chemicals were such successful poisons that there seemed to be no need to continue carrying out many of the old pest control practices, which previously had been a preventative habit—rotation, crop sanitation, encouragement of natural enemies, special cultivation practices, drainage of standing water for mosquito control, and similar operations. In some instances these practices were simply disregarded and discontinued.

Many students and researchers in the pest control disciplines became increasingly concerned with studying the killing efficiencies of chemicals. Research in this area was emphasized, often at the expense of research directed at gaining a better understanding of the biologies of the pests and their natural enemies. Thus the control of pests, which had always been considered a fundamentally ecological problem, began to assume the trappings of an offshoot of chemistry and engineering, sometimes involving little or no ecological understanding. This trend is perhaps best reflected in the subject matter of papers published in America's major applied entomological journal of the period, the *Journal of Economic Entomology*. Figure 4-10 shows that the number of papers in the *Journal* describing research in general biology (including records of pest incidence and damage, bionomics, ecology, and physiology) went down significantly during the period 1927–1952, while reports of research involving laboratory and field testing of insecticides began to clearly dominate the *Journal* after 1935.

The use of insecticides and other pesticides over this period became as normal and automatic to the grower as cultivating his field or sowing his seed. In the case of insect pests, he rarely bothered to see if the bugs were actually there in significant numbers—but simply sprayed according to a time schedule—e.g., weekly after seedling emergence until a week before harvest. It was an uncomplicated, easy-to-follow procedure, and growers regarded it as inexpensive and foolproof insurance against pest damage. And they were often urged on by pesticide company representatives, who had become the farmer's chief source of information about a wide range of pest problems. Unfortunately, problems with the heavy dependence on chemical control began to arise, and these problems were of an ecological–biological nature. They were ignored by most at first, and many were ignored for a long time; but eventually these problems

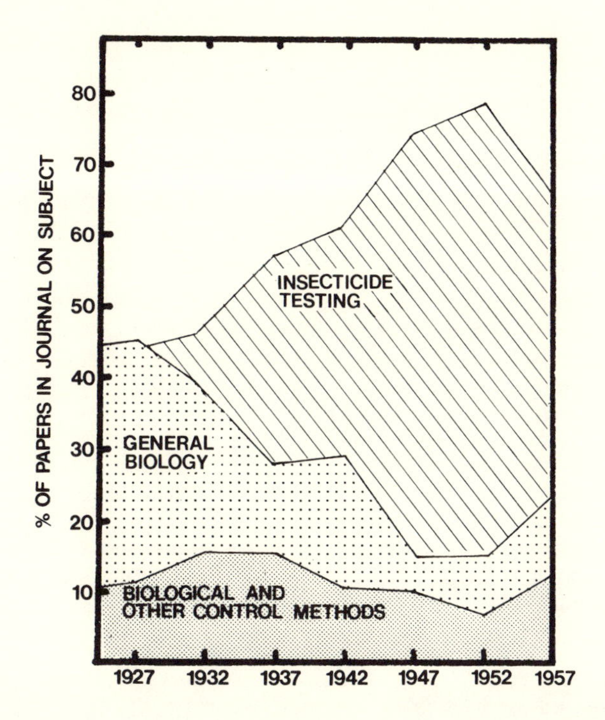

FIGURE 4-10. Trends in applied entomological research as reflected in the *Journal of Economic Entomology*, 1927–1957. Note how insect control research increasingly focused on insecticide testing and became less concerned with the biologies of the pests that were being controlled (data from Jones, 1973).

produced situations of such severity that they could no longer go unnoticed.

The earliest hint of impending disaster was the *development of resistance* to the killing power of the insecticides by some major pests. The first reported case of tolerance to DDT was in the house fly in Sweden in 1946. Within 20 years, some 224 species of insects and acarines had been recorded as resistant to one or more groups of insecticides: 127 agricultural and 97 pests of medical or veterinary importance. As of 1975, 75% of the most serious agricultural insect pests in California had developed resistance to at least one major insecticide, and in fact, a number had developed resistance to two or more materials (Table 4-2).

Insecticide resistance was not a new phenomenon, and it should not have been completely unexpected. Even before the arrival of the "miracle" insecticides in the 1940s, seven cases of resistance to the old-fashioned insecticides had been recognized: the resistance of the San Jose

TABLE 4-2
Pesticide Resistance in the Arthropod Pests Causing Over One Million Dollars Damage to California Agriculture in 1970[a]

Pest species	Types of pesticide for which resistance was demonstrated[b]	Resistance reported or suspected in California
Citrus red mite, *Panonychus citri*	DDT, OP, sulfur	×
European red mite, *Panonychus ulmi*	DDT, OP, sulfur	×
Pacific spider mite, *Tetranychus pacificus*	DDT, OP	×
Two spotted mite, *Tetranychus urticae*	DDT, OP, sulfur	×
Citrus thrips, *Scirtothrips citri*	DDT, cyclodienes, tartar emetic	×
Consperse stinkbug, *Euschistus conspersus*		
Lygus bug, *Lygus hesperus*	DDT, cyclodienes, OP	×
Pear psylla, *Psylla pyricola*	DDT, cyclodienes, OP	×
Cabbage aphid, *Brevicoryne brassicae*		
Citrus aphid, *Aphis citricola*		
Green peach aphid, *Myzus persicae*	DDT cyclodienes, OP, carbamates	×
California red scale, *Aonidiella aurantii*	OP, HCN	×
San Jose scale, *Quadraspidiotus perniciosus*	Lime sulfur	
Cotton bollworm, corn earworm, tomato fruitworm, *Heliothis zea*	DDT, cyclodienes, OP, carbamates	×
Beet armyworm, *Spodoptera exigua*	DDT, cyclodienes, OP, carbamates	
Cabbage looper, *Trichoplusia ni*	DDT, cyclodienes, OP, carbamates	×
Artichoke plum moth, *Platyptilla carduidactyla*	OP	×
Potato tubeworm, *Phthorimaea operculella*	DDT, cyclodienes, OP	
Pink bollworm, *Pectinophora gossypiella*	DDT, OP, carbamates	×
Peach twig borer, *Anarsia lineatella*	DDT, lead arsenate	×
Omnivorous leafroller, *Platynota stultana*		
Codling moth, *Laspeyresia pomonella*	DDT, OP, lead arsenate	×
Oriental fruit moth, *Grapholitha molesta*	DDT	
Cotton leaf perforator, *Bucculatrix thurberiella*	DDT, cyclodienes, OP, carbamates	×
Alfalfa weevil, *Hypera* sp.	Cyclodienes	×

[a] From Luck *et al.* (1977).
[b] DDT, DDT and relatives; OP, organophosphates; HCN, hydrogen cyanide.

scale in 1914 to lime–sulfur sprays; the resistance to California red scale in 1916, the black scale in 1916, and the citricola scale in 1938 to HCN fumigation; the resistance of codling moth larvae in 1928 to arsenical sprays; the resistance of screwworm larvae in 1942 to phenothiazine; and the resistance of the citrus thrips to tartar emetic–sucrose sprays in 1942. In contrast to the post-World War II situation, the use of these earlier materials was limited and the development of resistance was bothersome but not disastrous to the crops involved. Cultural methods or other insecticides were easily found to take the place of the ineffective materials. It is notable that in this 28-year period no insect developed resistance to more than one insecticide chemical. This phenomenon may be due partly to the nature of these materials and their modes of toxicity and partly to the fact that large populations of insects in the 1920s and 1930s were never exposed to the constant and repeated application of insecticides that became commonplace in the 1950s with the synthetic organic materials. The development of resistance to pesticide chemicals has not been limited to insect pests. While only a few cases have been reported, plant pathogens, weeds, and rodents have all developed strains resistant to chemicals applied for their control.

Development of resistance in a population of an organism is a common and logical evolutionary reaction to stress. Without a remarkable ability to evolve rapidly and adapt to sudden and often drastic changes in climate and habitat, insects would never have been able to dominate the animal kingdom as they have for many eons. The development of resistance is also known in the area of medicine, where certain strains of human disease organisms are no longer killed by medication (for instance, bacterial resistance to penicillin). When a large population of an organism is exposed to a certain kind of stress such as a toxic chemical, sometimes one or a few individuals may survive while the rest of the population is killed. This may be due to some physical factor (e.g., they were protected from exposure to the toxic chemical owing to sloppy application technique or for some other reason); however, survival may also be the result of one or more traits carried in the individuals' genetic makeup (their chromosomes) which somehow make them less susceptible to the toxin. Examples of such characteristics include the ability to manufacture detoxifying enzymes, behavioral mechanisms that prevent fatal exposure, a less permeable epidermis, or similar characters or combinations of characters. Since only individuals possessing these protective characters (or individuals that happen to escape because of some physical protection) will survive, it is easy to see that the next generation will contain a higher percentage of pesticide-resistant organisms. If every generation is exposed to the toxic chemical, soon only largely resistant individuals will constitute the population (Figure 4-11). Often a trait pro-

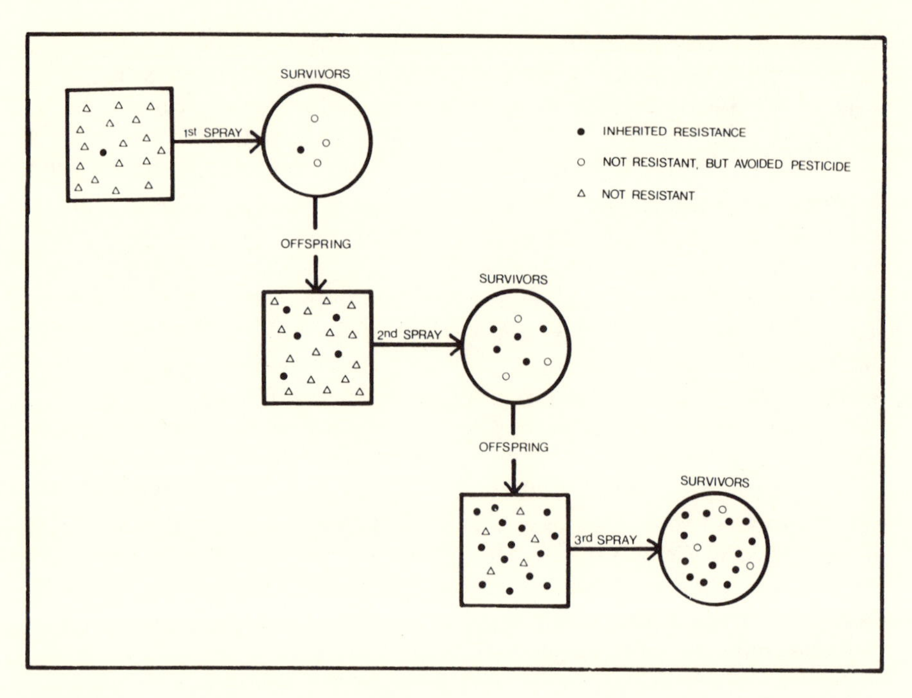

FIGURE 4-11. Schematic diagram of the increase in pesticide resistance in a repeatedly sprayed pest population over three generations.

viding resistance to one pesticide, such as a less permeable epidermis or an ability to detoxify a certain kind of poison, will allow the pest to tolerate another pesticide material as well, thus compounding the resistance problem.

Another problem growers began to notice was *target pest resurgence* (See Figure 4-12). After spraying with one of the modern insecticides to control a pest, growers noticed that its populations would sometimes drop drastically and then suddenly surge to higher levels than before. Pest resurgence occurred because the insecticides, as broad-spectrum poisons, killed natural enemies of the pest as well as the pest itself. Any natural enemies surviving the insecticide application would often starve to death since pest populations would be temporarily too low to provide adequate food; they would be forced to emigrate to other fields in search of food, or sometimes they would go into a reproductive lapse because of food shortage. The pest insects, on the other hand, would be able to do better than ever; their food source (the crop) would be readily available, often virtually inexhaustible, and now there would be no natural enemies to restrict or limit their population growth. Figure 4-13 shows such a case

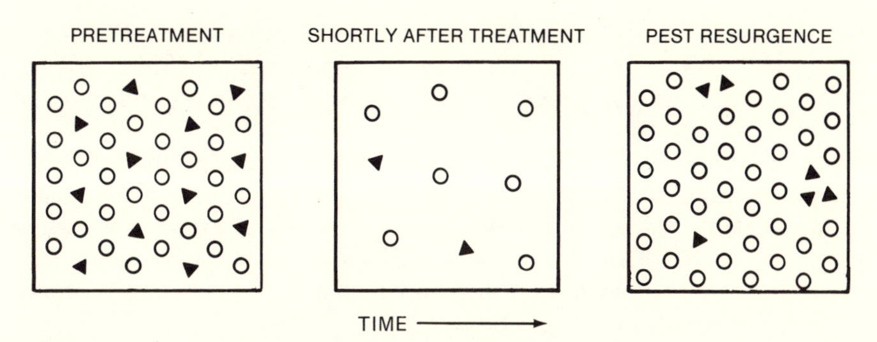

FIGURE 4-12. Target pest resurgence. Diagrammatic sketch of the influence of chemical treatment on natural enemy pest abundance and dispersion, and resulting pest resurgence. The squares represent a field or orchard immediately before, immediately after, and some time after treatment with an insecticide for control of a pest species (○). The immediate effect of the treatment is a strong reduction of the pest but an even greater destruction of its natural enemies (represented by ▲s). The resulting unfavorable ratio and dispersion of pest individuals to natural enemies permits a rapid resurgence of the former to damaging abundance (from Smith and van den Bosch, 1967).

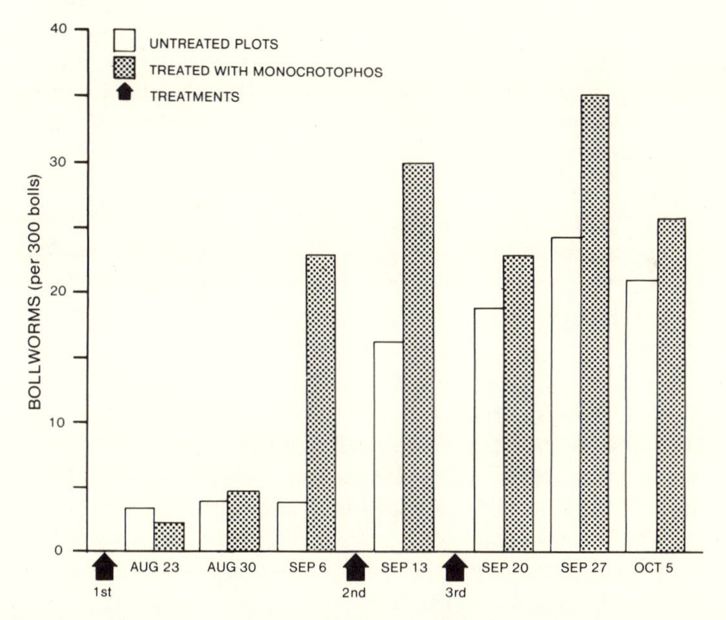

FIGURE 4-13. Target pest resurgence following applications of an insecticide for "pest control." In this experiment, plots treated with monocrotophos, an insecticide federally registered for bollworm control, suffered heavier bollworm infestations than untreated plots. Simultaneous samplings of predators revealed that the insecticide destroyed bollworm predators, which permitted resurgence of the pest. The data are from an experiment conducted at Dos Palos, California, in 1965 (from van den Bosch and Messenger, 1973).

of pest resurgence involving the bollworm in California cotton. Field sampling showed that the rapid increase in the population of the pest later in the season was indeed due to the destruction of predatory and parasitic insects.

The third type of problem engendered by dependence on the "miracle" insecticides is that of *induced secondary pest outbreak*. This occurs when a plant-feeding species, previously not a pest, suddenly erupts to damaging levels. This eruption is usually the result of the pesticides' destruction of natural enemies, which until then had kept the new pest under effective biological control (Figure 4-14).

A model case of secondary pest outbreak presented itself when DDT sprayed in California citrus orchards for control of other citrus pests caused a devastating outbreak of the cottony cushion scale. The cottony cushion scale, as discussed earlier in the chapter, had been kept under complete biological control since 1890 when the predaceous vedalia beetle and a parasitic fly were imported from Australia and established in California citrus orchards. These natural enemies exerted such an effective control that the cottony cushion scale had been almost forgotten. Their presence became painfully apparent, however, when the vedalia beetle proved particularly susceptible to DDT and the scale, released from nearly sixty years of biological control, again caused havoc in the citrus orchards where DDT had been used. It was not until DDT applications were adjusted and new populations of the vedalia beetle reestablished themselves in the sprayed orchards that the cottony cushion scale again ceased to be a pest.

The common reaction to these three repercussions from the use of the modern pesticides—(1) pest resurgence, (2) secondary pest outbreak, and (3) pest resistance—was an increase in pesticide use. When an insect developed resistance to a low dose of an insecticide, heavy doses would be applied until the pest could finally be killed, or another insecticide or a combination of several insecticides would be used. When a pesticide application resulted in target pest resurgence, the pesticide would be applied more and more frequently. And when a secondary pest outbreak occurred, the new "pest" would be treated like the original pest and extra applications (often involving additional materials) added to the spray schedule. The result of this increased use of pesticides was more pesticide resistance, more pest resurgence, and more secondary pest outbreak! This syndrome has been aptly termed "*the pesticide treadmill*"—once on it the farmer could not seem to get off. An excellent example of farmers being caught on the pesticide treadmill is discussed in the following chapter in the story of central American cotton.

A fourth problem resulting from the use of the "miracle" insecticides has been *environmental contamination*. The potential environmental haz-

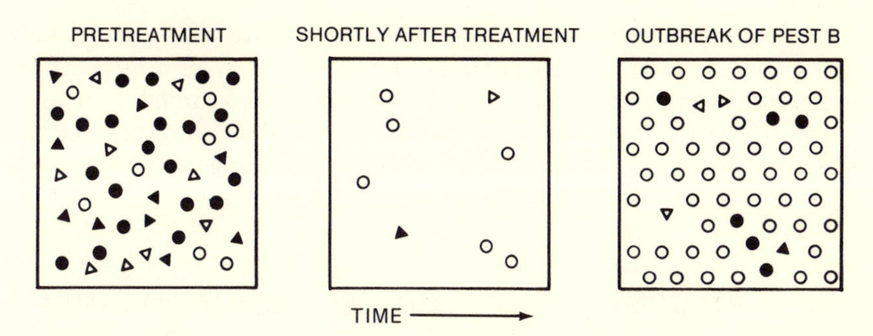

FIGURE 4-14. Secondary pest outbreak. Diagrammatic sketch of the influence of a chemical treatment on natural enemy, pest abundance, and dispersion with resulting secondary pest outbreak. The squares represent a field or orchard immediately before, immediately after, and some time after treatment with an insecticide for control of pest A (●). The chemical treatment effectively reduces pest A as well as its natural enemy (▲), but has little or no effect on pest B (○). Subsequently, because of its release from predation by predator B (△), pest B flares to damaging abundance (from Smith and van den Bosch, 1967).

ards posed by the use of such broadly lethal zooicides had been pointed out right after the introduction of DDT by a tiny minority, but these dissenters received little public attention until the publication of *Silent Spring* by Rachel Carson in 1962. People soon discovered that these poisons, especially the chlorinated hydrocarbons like DDT, were everywhere in the environment—in Antarctic penguins, boreal frogs, fish in the depths of oceans, the lowliest decomposer organisms, and the milk of human mothers. Pesticides were being widely applied and then drifted via wind and water to places remote from the areas of application. Agricultural workers, pest control operators, and other people exposed to various insecticides, especially organophosphates, became the victims of both acute and chronic poisoning.

The United States banned most uses of DDT in 1972, and subsequently has severely restricted or banned the use of aldrin–dieldrin, endrin, heptachlor, DBCP, and chlordane. But the problems of environmental contamination have not disappeared with the banning of a few chemicals. In many cases, agriculturalists, foresters, pest control operators, and other pesticide users have simply substituted new products, and with these new products, new problems have resulted. The 1976 story of the poisoning of industrial workers and the crippling of a major fishing industry by the insecticide Kepone points out succinctly how close to home environmental contamination must get before people will take the issue seriously. In Chapter 5, several case histories of environmental contamination by pesticides are discussed in more detail. Clearly, the only way to curtail such contamination is to make much more discriminating use of these environmental poisons. Well-designed and carefully executed integrated pest management programs, relying extensively on cultural and biological controls, can contribute greatly to minimizing pesticide abuses.

THE DEVELOPMENT OF INTEGRATED PEST MANAGEMENT

The development of integrated pest management has been the most recent chapter in the history of pest control. In integrated pest management, various combinations of methods are utilized in a compatible manner to obtain the best control with the least disruption of the environment. Although many of the cultural, physical and biological control methods worked out in the first third of the twentieth century are utilized, IPM is not, as some might think, a return to pre-World War II pest control. While many good techniques for individual pest problems were worked out during that period, these techniques were independently developed and were rarely coordinated into pest management programs that evaluated the effects of two or more pest management operations on each other. Although early methods often were the result of an admirably sound biological knowledge of pest life cycles and were directed at the pest's "weak points," few of the methods recognized the importance of assessing population numbers of both pest and natural enemy populations to predict future population trends and determine if pest control action was actually needed. This is the key to integrated pest management.

Although the concept of integrated pest management has only been popularly accepted for the last ten to fifteen years, the roots are much older. In the late 1940s, Ray F. Smith and others suggested the need for supervised control specialists who would carry out routine field monitoring of pest populations and their natural enemies and who would prescribe to the grower what, if any, control action was needed. This suggestion has been significantly implemented only in the last decade and a half.

Agricultural entomologists were at the forefront of the development of integrated pest management. Perhaps because the problems of pest resistance, pest resurgence, and secondary pest outbreak have been most severe among insect pests, it was a group of entomologists who first elaborated the concept of economic levels and thresholds and the concept of integrated control itself. Over the last thirty years, entomologists have also perfected several new control tools compatible with the integrated pest management concept and minimally disruptive to ecosystems. These tools include the use of insect pathogens for pest control, the use of in-

TABLE 4-3
Major Events in the History of Pest Control

Date	Event
400,000,000 B.C.	First land plants
350,000,000 B.C.	First insects
250,000 B.C.	Appearance of *Homo sapiens*
12,000 B.C.	First records of insects in human society
8,000 B.C.	Beginnings of agriculture
4,700 B.C.	Silkworm culture in China
2,500 B.C.	First records of insecticides
1,500 B.C.	First descriptions of insect pests
950 B.C.	First descriptions of cultural controls (burning)
300 A.D.	First record of use of biological controls (predatory ants used in citrus orchards in China)
1650–1780	Burgeoning of insect descriptions (after Linnaeus) and biological discoveries in Renaissance
1732	Farmers first begin to grow crops in rows to facilitate weed removal
1750–1880	Agricultural revolution in Europe
Early 1800s	Appearance of first books and papers devoted entirely to pest control
1840s	Potato blight in Ireland (no controls available to curb disaster)
1870–1890	Grape phylloxera and powdery mildew controlled in French wine country (introduction of Bordeaux mixture and Paris Green; use of resistant rootstalks and grafting)
1880	First commercial pesticide spraying machine
1888	First major biological importation success (vedalia beetle for control of cottony cushion scale)
1890s	Introduction of lead arsenate for insect control
1896	Recognition of arthropods as vectors of human disease
1896	First selective herbicide (iron sulfate)
1901	First successful biological control of a weed (lantana in Hawaii)
1899–1909	Development of strains of cotton, cowpeas, and watermelon resistant to *Fusarium* wilt (first breeding program)
1912	U.S. Plant Quarantine Act
1915	Control of disease-vectoring mosquitoes allowed completion of Panama Canal

(*Continued*)

TABLE 4-3 (*Continued*)

Date	Event
1921	First aircraft spray (in Ohio for catalpa sphinx)
1929	First area-wide eradication of an insect pest (Mediterranean fruit fly in Florida)
1930s	Introduction of synthetic organic compounds for plant pathogen control
1939	Recognition of insecticidal properties of DDT
1940	Use of milky disease to control Japanese beetle (first successful use of insect pathogen for control)
1940s	Organophosphates developed in Germany, carbamates in Switzerland
1942	First successful breeding program for insect pest resistance in crop plants (release of wheat strain resistant to Hessian fly)
1944	First hormone-based herbicide (2,4-D)
1946	First report of insect resistance to DDT (housefly in Sweden)
1950s, 1960s, and 1970s	Widespread development of resistance to DDT and other pesticides
1950s	First applications of systems analysis to crop pest control
1959	Introduction of concepts of economic thresholds, economic levels, and integrated control
1960	First insect sex pheromone isolated, identified, and synthesized (gypsy moth)
1962	Rachel Carson's *Silent Spring*
1972	Banning of DDT in United States

secticides that selectively kill pests with minimum negative effects on beneficial organisms, the use of insect pheromones (especially sex attractants, which have been particularly useful in sampling populations), and the use of genetic manipulation (e.g., host resistance to pests and release of sterile males). Methods of monitoring populations have vastly improved, as has the entomologist's ability to understand the factors that contribute to pest outbreaks through the development of computerized models that can simultaneously consider far more variables in the managed ecosystem than can the unaided human brain.

Plant pathologists, weed scientists, rodent control specialists, wildlife managers, and many other ecosystem management and pest control specialists have begun jointly to develop complete integrated pest management systems for the ecosystems that they manage. Examples of these developments are described in the chapters that follow. Major events in the history of pest control are summarized in Table 4-3.

CHAPTER 5

The Cost of Pest Control
Economic, Social, and Environmental

Pest control and other resource management actions are ecosystem manipulations designed to maintain resource quality, preserve or improve human health and comfort, or enhance the production of food and fiber. Their employment requires a workforce, materials, energy, and the modification of the environment. Their implementation represents a conscious decision to allocate materials, energy, and people to the production of certain benefits and to forego other possible priorities. As such, these actions have economic, social, and environmental ramifications for all of society. People involved in the design and implementation of resource management programs must necessarily concern themselves with the direct (or private) short-term economic benefits of these programs; but even so, these persons need to be aware of the far-reaching effects of their actions on the general public.

THE PRIVATE ECONOMICS OF PEST MANAGEMENT

How does a grower, forester, or other resource manager make the decision to use a pest control chemical or to employ some other management action? Sometimes the decision is never consciously made; a material is used or an action taken out of habit, on the advice of a manufacturer's representative, or because it appeared to be a potentially rewarding option or perhaps the "only option."

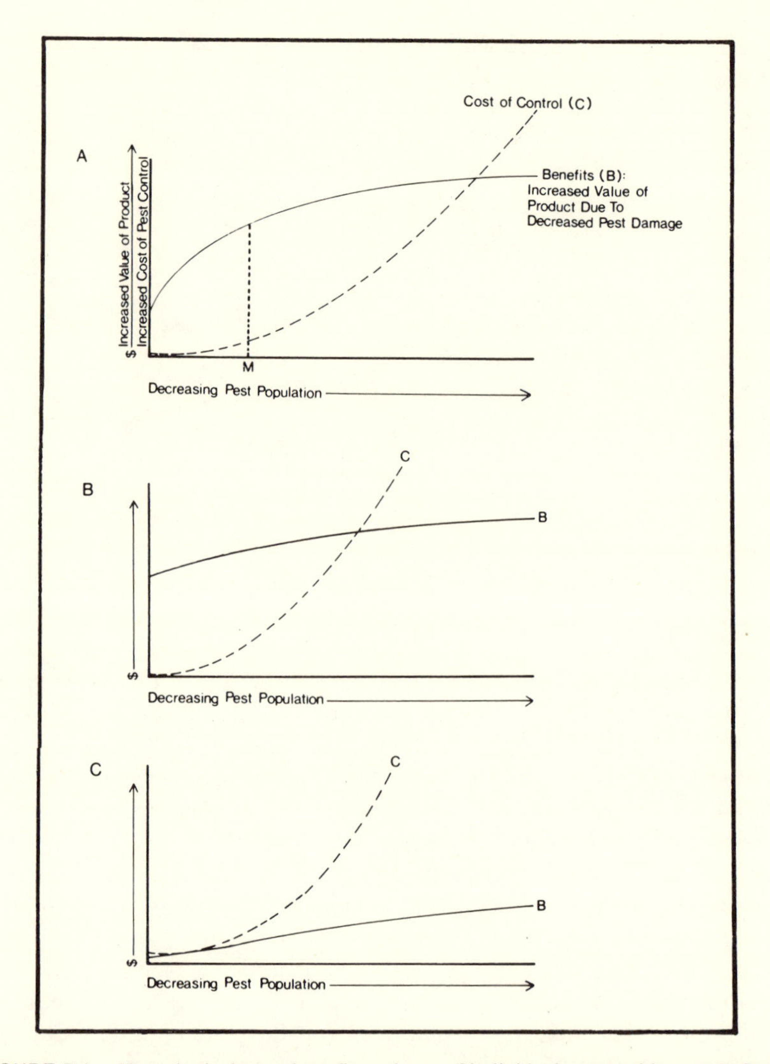

FIGURE 5-1. Hypothetical cost–benefit analyses of individual pest problems. (A) Optimal pest control strategy (i.e., largest value of $B - C$) is at M. (B) Optimal pest control strategy is to spend no money on pest control (i.e., largest value of $B - C$ occurs when cost of control is zero). (C) Pest population is below economic levels; cost of control always exceeds benefits (i.e., pest control action cannot be justified).

The more thoughtful resource manager will evaluate the populations of pest and beneficial organisms and, with the use of control action thresholds (discussed in Chapters 3 and 7), determine whether the cost of pest control can be justified economically. The resource manager who wants to truly get the most from his pest control investment will also make a more detailed cost–benefit analysis to determine the optimum pest control action. Figure 5-1 shows hypothetical cost–benefit analyses of individual pest problems.

In theory, up to the point where all pest individuals have been destroyed, as more effort (=money) is expended on the control of a pest, damage drops and the value of the product increases (Figure 5-1A). If the pest population is brought down to zero, damage will be totally eliminated, and the value of the product will be increased to the maximum possible owing to pest control action. However, attainment of the maximum increased value of product is usually not worth the cost of the necessary pest destruction operation (e.g., daily pesticide applications or hand removal of each pest individual as it appears). Correspondingly, the economically minded resource manager seeks to optimize the return from his pest control investment rather than maximize the ultimate value of this crop.

According to cost–benefit analysis, the net profit to the producer from a pest control action is equal to the benefits (B) derived from that action (i.e., the increased value of the resource) minus the cost (C) of that control action; thus the net profit is $B - C$. (Note: in using a simple economic threshold analysis, economic levels would be reached anytime $B - C$ is more than zero.) In the hypothetical cost–benefit analysis illustrated in Figure 5-1A, the largest value for $B - C$ is at M. Thus, if the producer reduces the pest damage to M at the indicated cost, he will optimize his pest control investment. Sometimes the greatest cost–benefit difference can be derived by taking no action at all, as in Figure 5-1B. Although in this case the producer can reduce pest damage by investing money in a control action and still retain a net profit, the profit is not as large as if he had taken no action at all. Figure 5-1C describes a third case where treatment cannot be justified economically in any instance; in other words, the orginal pest population never reaches damage levels that justify control costs.

A serious problem with the simple analysis provided by either present-day economic or control action thresholds or this kind of cost–benefit analysis is the limit of their scope. Managed ecosystems are dynamic environments composed of and affected by hundreds of interacting variables (Figure 5-2), including weather, control actions, cultural practices, and living organisms. Yet with current action thresholds the resource manager can consider only one at a time: the effect of that pest on the value of only one resource or the effect of the pest control action on only the single pest. He has not considered the effect of the control action on other pests (e.g., such action might produce a secondary pest outbreak requiring even more control expenditure, or it might suppress other injurious species in the treated area) or the possible direct effects the action

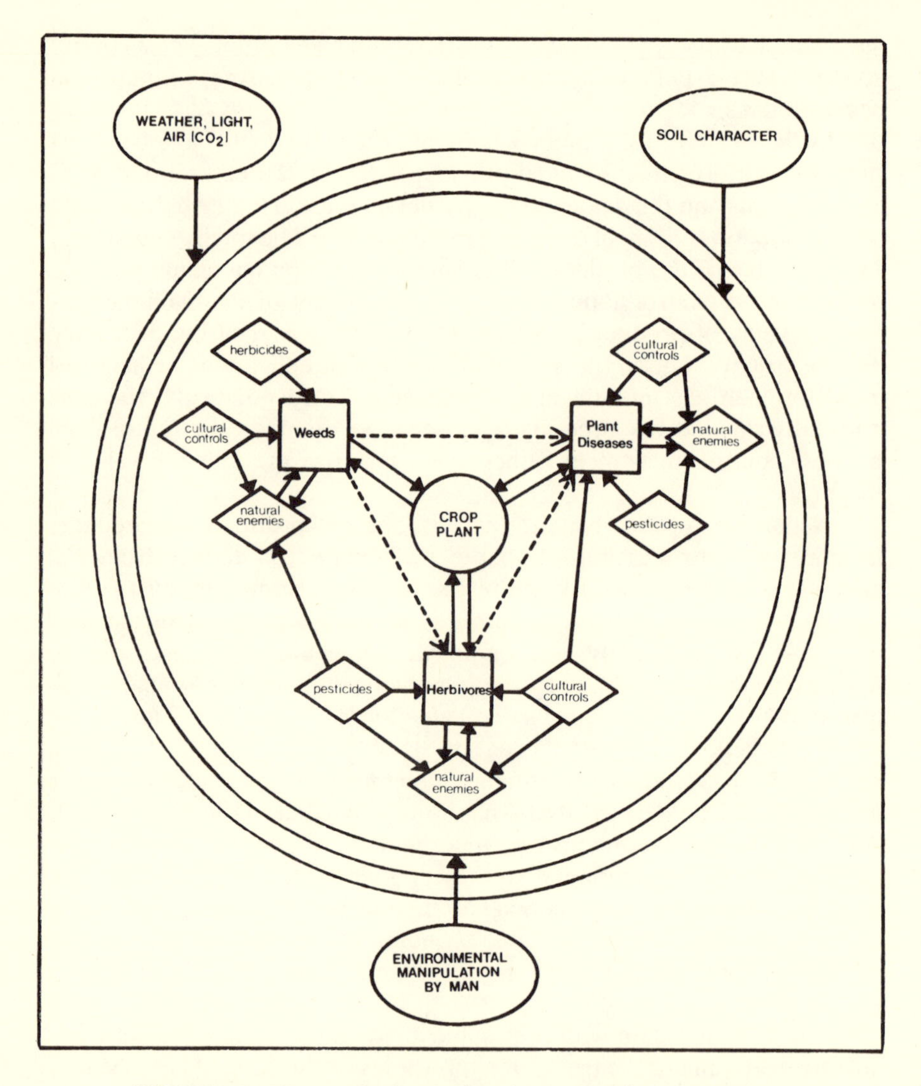

FIGURE 5-2. Some major interactions in a managed ecosystem.

might have on the crop itself (e.g., a pest control chemical might be damaging to leaves (phytotoxic) and thus reduce yield, or it might leave a toxic residue that would make the crop unmarketable). Also, the analysis unrealistically assumes that the pest control action will have an equal effect regardless of weather, other cultural practices, time of the year, or other stresses on the resource.

Researchers are just beginning to evaluate the interactions and effects of cultural practices, fertilizing, pest control, and other resource management actions on each other and on other parts of the resource ecosystem. This research is being done with the use of computer models that simulate parts of the resource ecosystem and the manipulations managers may make in that environment. With the knowledge accrued from this research, cost–benefit and economic (and aesthetic) threshold estimations will be expanded and adjusted to suit different types of situations. Meanwhile, the resource manager and pest management specialist must depend on the crude estimations now available plus their own experience and knowledge of pest biology, the ecosystem, and seasonal changes to make management decisions that optimize resource production and maintenance. For an amazingly high percentage of our agricultural pests, we have not even developed rough estimates of their impact on crop yield.

There are many long-term effects a resource manager must take into account. One is the potential buildup of resistance to a pesticide in a pest after heavy reliance on chemical control; if that resistance causes the pest to become more difficult to control or even uncontrollable in the future, the producer has certainly paid a dear price for pest control. Another consideration should be the actual value of the resource at harvest time. For instance, it is conceivable that if most of the growers of an agricultural crop suffer a significant yield loss because of a pest then the supply of that crop will be limited, and if the demand remains somewhat stable, the per value of the crop may go up. Thus if the grower based on the cost–benefit analysis on last year's value of the crop, he may have underestimated the amount of money actually available for pest control. Growers and other resource managers should also include in their cost–benefit analysis a consideration of the possible hazard to pollinators, legal risks arising from drift of toxic substance to neighboring properties, and other possible financially adverse effects of pesticide application.

The management of pests in resources such as city parks, recreation areas, and home gardens, which are not maintained for an economic return, cannot be evaluated on strictly economic terms. Ideally, management in these areas focuses on long-term maintenance of a balanced and healthy ecosystem. Thus pest management programs in these ecosystems should be among the first to weigh heavily the long-term social and environmental consequences of control actions. However, until now, resource management in every category has consistently failed to take this broader ecosystem view, and, as will be illustrated in the rest of this chapter, this failure has often led to serious repercussions.

CONFLICTS OF INTEREST IN ECOSYSTEM MANAGEMENT

Failure to view pest control and other management actions as manipulations that have effects on many parts of the resource ecosystem

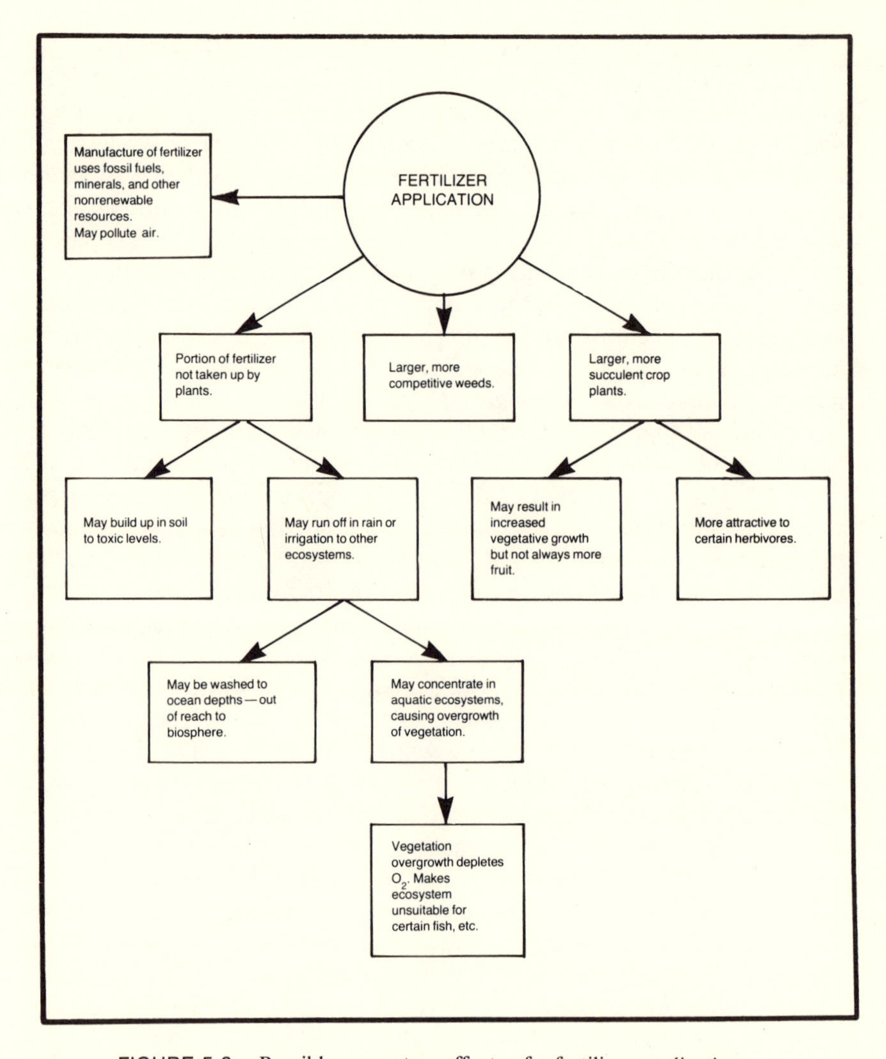

FIGURE 5-3. Possible ecosystem effects of a fertilizer application.

and the surrounding environment continues to cause problems in every area of resource mangement. A look at the broader ramifications of four common resource management actions will illustrate this. In each case the resource manager is satisfied that he has solved his immediate problem but may not be aware that his actions have created several greater problems.

1. Fertilizer (Figure 5-3)

A grower decides to improve his yield by applying a large amount of an inorganic N, P, and K fertilizer to his irrigated field. He will probably be pleased with the results; his crop plants are larger and more succulent and have grown rapidly. Yet as a by-product of his heavy fertilizer application he has managed to produce larger, more rapidly growing, and more competitive weeds; these will require extra attention. His succulent, vigorous plants may also have attracted more insect pests. Entomologists working in cotton, tobacco, and a number of other crops have shown that many adult insects preferentially oviposit in succulent growth. Cotton researchers have also noted in crop's where it is the fruit that is harvested, the added vegetative growth induced by overfertilizing can actually be counterproductive.

What are the long-term effects of this added fertilizer? If drainage is not ideal, there may be a buildup of salts in the soil, which could reduce the productivity of the field and sometimes even concentrate them to such toxic levels that the land becomes virtually nonarable.

Fertilizer use has effects on the surrounding ecosystems as well. Irrigation and rain-flushed runoff of these nutrients into streams, rivers, and other bodies of water may cause an overgrowth of certain aquatic vegetation. In extreme instances, giant clumps of algae and other aquatic plants may clog water flow, rendering it turbid, nonpotable, oxygen-deficient, and uninhabitable for fish and many other organisms. Excessive drainage of fertilizer-derived salts into a river system can so increase the salinity of the river that the waters in its lower reaches are no longer useful for irrigation. A case in point is the Colorado River, which picks up so much fertilizer-derived salt in Arizona and California that the utility of its waters for Mexican farm lands is now threatened. A considerable fraction of the nutrients washed off fields and forests into streams and rivers ends up in the bottom of the ocean, out of reach of organisms and removed from the material cycles of the biosphere.

2. Forest Herbicide Application (Figure 5-4)

A forester applies a phenoxy herbicide such as 2,4-D, silvex, or 2,4,5-T to a recently harvested forest area to keep broad-leaved undergrowth from shading out newly planted conifer seedlings. The herbicide application is a success; the broad-leaved plants are killed or severely stunted, and the conifers get the head start that they need to skip a few successional stages and rapidly dominate the forest.

But what are the other effects of the kill-off of all these broad-leaved plants? First of all, those animals that depend on broad-leaved plants for

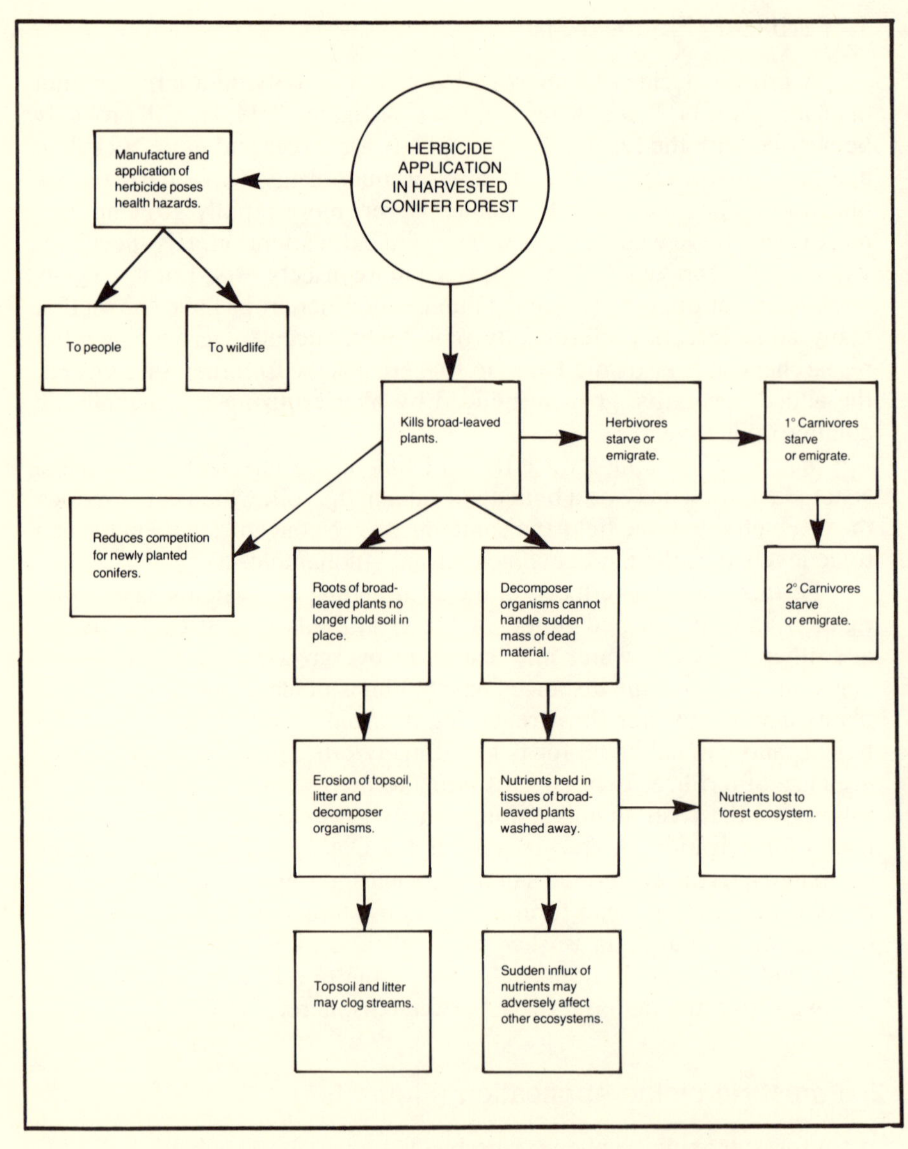

FIGURE 5-4. Possible ecosystem effects of a herbicide application.

food starve or emigrate. And the animals that feed on those herbivores suffer similarly. Second, erosion sets in. The leaf litter and roots of the small and medium-sized broad-leaved plants keep the topsoil in place during the periodic flooding of the forest. These plants are also important in holding in soil moisture and keeping wind erosion at a minimum.

Broad-leaved plants play a vital role in the retention of the forest ecosystem nutrients, particularly after the taller conifers have been harvested. Within their tissues they hold large quantities of essential nutrients. In a natural situation this store of nutrients can be cycled back into the forest ecosystem through the reducing action of detritus and decomposer organisms as the plants die and are replaced one by one. By preventing erosion and holding the topsoil in place, these broad-leaved plants also preserve the store of nutrients and the decomposer organisms contained within the top layers of soil and litter.

The preservation of nutrients and prevention of erosion is important in the maintainance of surrounding ecosystems as well. As discussed earlier in the fertilizer example, excess nutrients washed into aquatic ecosystems can result in overgrowth of certain vegetation, reduction of many aquatic animal and plant species, and the permanent loss of nutrients from the biosphere. Also, excess runoff of topsoil and other particulate material can clog creeks and rivers and fill reservoirs.

The herbicides 2,4,5-T and silvex have another effect, which, until recently, the forester has not fully considered. These two herbicides carry a contaminant, TCDD (a dioxin), that is highly toxic in minuscule quantities to a wide range of animals, including humans, and also has suspected teratogenic impacts on human babies. These serious hazards have led many to believe that the benefits of these herbicides are not worth the risk to society, especially to workers involved in their manufacture and application and residents of areas around manufacturing facilities and application sites.

3. Fungicide

An apple farmer previously plagued by apple scab (a fungus disease) sprays his trees with a large dose of copper–sulfur compound annually. He appears to have met his goal; his trees give crops with no sign of apple scab year after year.

Yet he may have created new problems. Sulfur-containing compounds may have a negative effect on populations of natural enemies (especially parasites) of some insect pests and spider mites, and the grower may consequently experience an increase in insect or mite populations. These new pests may have to be treated with insecticides; this, in turn, may induce a pesticide treadmill. Sulfur and copper sulfate exposure are also major cause of eye and skin ailments among pesticide applicators and other agricultural workers.

Copper–sulfur compounds produce stable residues which remain in the soil. The compounds may have extensive effects on the soil flora (especially fungi) and fauna, all of which are important in breaking up and decomposing litter and debris and in transforming nutrients from dead organic matter into a form that can be taken up and utilized by higher plants.

An unusual case from Great Britain will illustrate this point. An apple orchard, heavily sprayed year after year with copper compounds, showed a marked paucity of soil-inhabiting organisms. The buildup of dead leaves and litter in this orchard was astounding. Over the winter no decomposition took place, whereas in neighboring orchards the litter was rapidly broken down. The accumulation of copper had killed a wide array of decomposer organisms in the litter layer and near the surface of the soil. To compound this type of problem, such persistent poisons do not biodegrade and stay in the orchard virtually forever. However, it is important to point out that there is no evidence of these problems occurring with EPA labeled rates of copper sulfate.

More modern organic fungicides are less persistent than copper sulfate compounds and are apt to have less severe effects on the surrounding ecosystem. Copper sulfate compounds are not ancient history, however. Bordeaux mixture (copper sulfate, quicklime, and water) is still the most widely used fungicide in the world.

4. Insecticide Application—The Central American Cotton Disaster

In modern times, the ill-advised use of insecticides has perhaps engendered as many cases of economic loss, environmental contamination and societal calamity as any other resource management action. One of the best studied and most tragic stories has been documented by the Central American Research Institute for Industry (ICAITI) in their 1976 report on the environmental and economic consequences of pesticide use in central American cotton production. The Central American countries involved in this episode were El Salvador, Guatemala, Honduras, and Nicaragua. In the fifth Central American country, Costa Rica, cotton is not an important crop.

The story began in 1950 with the mass introduction of modern farm machinery and the organosynthetic insecticides. There were two economic pests in cotton in Central America at that time, the boll weevil (*Anthonomus grandis*) and the leafworm (*Alabama argillacea*). At first the most commonly used insecticides were the organochlorines DDT, BHC, and toxaphene. A little later organophosphate insecticides were added to the list. Initially, less than five insecticide applications were made per season. From the beginning these new insecticides seemed a

TABLE 5-1
Pesticide Treadmill in Central American Cotton

	1950	1955	1960s
Number of pesticide applications	0–few	8–10	28
Major pest	Boll weevil Leafworm	Boll weevil Leafworm Bollworm Cotton aphid False pink bollworm	Boll weevil Leafworm Bollworm Armyworm (2 species) Whitefly Cabbage looper Plant bug

great success—almost miraculous. The per hectare yield increased from 1550 kg of cotton (pre-1950) to 2270 kg. Anticipating even greater yields, growers hastened to get more hectares into cotton production.

By 1955, however, a few unforeseen problems had emerged (Table 5-1). Heavy dependence on insecticides resulted in secondary pest outbreaks. Three insect species previously unknown as pests became economically damaging: the bollworm (*Heliothis* sp.), an aphid (*Aphis gossypii*), and the false pink bollworm (*Sacadodes pyralis*). Additional insecticide treatments were made for these new pests. The new insecticide applications decreased the seriousness of the false pink bollworm and the aphid as pests but created in their place four other major insect problems: two genera of armyworms (*Prodenia* spp., *Spodoptera* sp.), a white fly (*Bemisia tabaci*), and the cabbage looper (*Trichoplusia ni*). Later, a plant bug (*Creontiades signatum*) joined the lineup of induced pests. By the 1960s the average number of insecticide treatments had leaped to 28 a year. Treatments were applied on a strict calendar schedule without regard to pest population numbers.

The situation reflected the typical "pesticide treadmill" syndrome: insecticides causing target pest resurgence and secondary pest outbreaks, more insecticide applications creating problems of pest resistance to the chemicals, and the grower continuing to make heavier, more frequent applications. Pest control expenditures skyrocketed until they accounted for 50% of all production costs in Central American cotton.

Encouragement from foreign pesticide manufacturers had made it easy for the Central American cotton growers to become victims of the pesticide treadmill. Since these countries did not have the stringent safety regulations for pesticide chemical use that countries in North America and Europe had, large manufacturing firms saw Central America as an ideal place to test their new materials. Foreign pesticide manufacturers showered growers in these countries with scores of pesticidal chemicals and lots of advice. The Central American growers, under the mistaken assumption that the key to combatting pest resistance was mixing the different materials, welcomed the new chemicals and the free "technical" advice.

But despite the chemical "cocktails" and all the free advice, the situation continued to deteriorate and by the early 1970s looked grim. Pest control costs were beginning to drive growers out of business. Yields were declining. There were even reports that excessive pesticide use was killing the cotton plants themselves. The cost of treating the increasing number of poisoned farmworkers was another concern for some producers as well as government health authorities. Even in 1972, when pesticide use was being curtailed, 6063 workers in Central America were treated for pesticide poisoning; undoubtedly many others went untreated.

Cotton pest control caused other agricultural industries to suffer losses as well. Frequent blanketing of the countryside with insecticides for cotton insect control induced insecticide resistance in pests of other crops. The most serious case occurred in corn where the insect transmitter of corn stunt virus could no longer be controlled; the cultivation of corn in certain cotton-growing areas of Nicaragua became impossible. Pesticide contamination of air, water, pasture plants, and cottonseed-based feed concentrates caused the buildup of high pesticide residues in cattle. As a result, beef was rendered unfit for export and milk unfit for domestic consumption.

Environmental contamination was rampant. High pesticide residues were found in fish, earthworms, amphibians, reptiles, birds, shellfish, and edible green plants. These included many important food items.

Public health officials charged with malarial control programs became alarmed with the rapid increase of insecticide resistance in *Anopheles albimanus*, the mosquito that vectors the malaria parasite in these cotton-growing areas. Although initially the incidence of malaria was lower in cotton-growing areas (probably owing mostly to the reduction of mosquito populations by pesticide sprays for cotton insects), health officials feared this trend would quickly reverse itself with the emergence of insecticide-tolerant mosquitoes. As discussed in the previous chapter, repeated pesticide application will give a great survival advantage to pesticide-tolerant individuals; these strains multiply and soon predominate. As Table 5-2 indicates, the heavy selection pressure provided by constant insecticide applications in certain Central American cotton-growing areas was enough to encourage the growth of such insecticide-tolerant mosquito populations.

Although environmental contamination and hazards to human health were important contributing factors, it was the economic disaster in the

TABLE 5-2
Insecticide Resistance in *Anopheles albimanus* (1974)[a]

	DDT (Guatemala)	Propoxur (OMS-33)[b] (El Salvador)
Cotton-growing areas	79%	35.4%
Non-cotton-growing areas	53.6%	6.94%

[a] From ICAITI (1976).
[b] OMS-33 was the insecticide recommended for mosquito control in Central America in 1974.

cotton industry which finally led to a change in pest management policy. In the early 1970s the Central American countries began to work out integrated pest management programs for cotton involving monitoring, cultural controls, preservation of natural enemies, and reduced insecticide use. Within a few years, pesticide use had dropped 30–40%, and the cotton industry was on the road to recovery. However, environmental contamination and the consequences of chronic human exposure to pesticides will remain for years as grim reminders of the damage unwise pest control can cause, not only to the managed ecosystem but also to the surrounding ecosystems and to the people living and working nearby.

THE SOCIETAL IMPLICATIONS OF PEST MANAGEMENT DECISIONS

Many of the problems that have resulted from pest control action in the past can be seen as products of what appears to be a basic conflict between the interests of the resource manager and the broader interests of the general public. The resource manager, striving to make a profit or keep the budget in the black, generally bases his decisions entirely on the short-term economic gains for himself (or his employer) and views those decisions as personal, private, and no one else's business. However, pest management decisions may well have broader economic effects on society, and may affect environmental quality or create hazards to human health.

Economic Effects

Pest management decisions may be of broader interest to society because of their effect on price of the resource to the consumer, the allocation of materials, energy, and labor to the production of the resource, or the inhibition or stimulation of other industries and resources by the management practices or production of this resource.

Ideally, if pest damage is reduced, yields increased, and pest control costs minimized by good pest management, the consumer benefits by lower prices for food, fiber, and other resources. The economic advantage of such good pest management is obvious, and if truly passed on to the consumer, benefits all of society. Unfortunately, in reality, such benefits are rarely passed on to the consumer.

If a pest management program is able to reduce the amounts of energy, materials, and/or labor required to produce a resource, then these materials, energy, and labor become available to meet other of society's needs. This can be a benefit, as in the reallocation of energy in an energy-short country, or it can be a burden, as is often the case with a reduction in the required amount of labor. For instance, if a more efficient resource management system diminishes the need for labor and forces a surplus of unskilled workers onto an already saturated labor market, many of these newly unemployed workers will have to seek support from governmental welfare agencies at a cost to all of society.

The economic effects of the management practices of one resource manager on other industries is more far-reaching than many realize. First, there may be benefits for neighboring resources. For instance, if a grower is able to suppress corn earworms (another name for the cotton bollworm) in his corn field early in the season, he might save his cotton-growing neighbor the cost of spraying for this pest later in the season, when it might have otherwise migrated into the cotton.

Of course, pesticide usage may hinder the production of other resources as well. For example, if a farmer growing wheat accidentally douses part of his neighbor's cabbage field with a broad-leaf herbicide, it will more than likely reduce his neighbor's cabbage yield. Drift of pesticides from one field to another may kill natural enemies in nontarget crops, disrupting biological control of pests; or it may help increase insecticide resistance in neighboring insect populations. Pesticide applications destroy hundreds of thousands of honey bee colonies in the U.S. every year (see Table 5-3 for California figures), inflicting serious economic losses both on beekeepers and the growers whose crop production depends on pollination by bees (e.g., almonds, apples, apricots, melons, alfalfa seed). The insecticides may also kill wild pollinators which help in overall crop production.

Other industries have been affected by pesticide drift and runoff as well. In 1963, the impact of pesticides on the fishing industry became apparent with the near shutdown of Louisiana commerical fishing owing to fish poisoning caused by insecticide runoff into the Mississippi from agricultural fields and industrial waste. Fish kills due to pesticide poi-

TABLE 5-3
Bee Colony Loss to Pesticide and Other Causes in California[a]

Year	Colonies of bees	Loss to pesticides	Other losses	Total loss
1969	537,000	82,000	117,000	199,000
1970	521,000	89,000	70,000	159,000
1971	511,000	76,000	32,000	108,000
1972	500,000	40,000	30,000	70,000
1973	500,000	36,000	31,000	67,000
1974	500,000	54,000	33,000	87,000
1975	500,000	31,000	28,000	59,000
1976	525,000	47,000	41,000	88,000

[a] From the California Department of Food and Agriculture.

soning were common in the decade that followed and continue today. In 1976 both the commerical and sport fishing industries were shut down in the Chesapeake Bay due to industrial pollution from a company manufacturing the insecticide Kepone®. Consumption of fish caught in Lake Ontario was prohibited in 1976 when that lake was discovered to be contaminated with industrial wastes containing the insecticide Mirex. While only the most devastating disasters made the news, less severe losses have been inflicted on fishing industries all across the North American continent as a result of drift and runoff of pesticides applied for agricultural and other pest control uses.

Other industries directly affected by pest management practices include the industries that manufacture and market pest management equipment and materials and those that apply them (e.g., aircraft applicators). Their interest, of course, is to remain in business and sell as much of their products or services as possible. Understandably, their interests and the interests of sound pest management based on minimal pesticide use may collide head-on.

Although industry depends on resource managers to implement strategies that require their products, resource managers are also dependent on industry to supply them with the kind of pest control materials with which they can effectively manage pests. Thus, the exchange between pest control practitioners and manufacturers is two-way. For instance, implementation of integrated control of insect pests could be greatly enhanced if selective insecticides were available that kill certain pest species but not their natural enemies. Such materials have been found and a few, such as the microbial insecticides and insect growth regulators, are commercially available; but since they kill only a narrow range of pest species in a limited number of crop situations, manufacturers have often not felt that the production and development costs could be justified by such a small market. Pesticide manufacturers have another large influence on pest control practices. Research has shown that agricultural producers rely heavily on advice from pesticide company representatives (i.e., salesmen–fieldmen)—despite their obvious economic bias toward recommending the use of pesticides—for information when making pest control decisions. It is hoped that in the future this source of information will be balanced by the growing industry of independent pest management specialists (see Chapter 9).

ENVIRONMENTAL QUALITY

We have seen a few examples of the effect of pest control actions on environmental quality in the discussion of economic losses to the fishing, apiculture, and agriculture industries resulting from pesticide contamination. However, most cases of environmental contamination—such as injury to or destruction of native plants and wildlife—do not have obvious and immediate economic effects and, unfortunately, are therefore often ignored, ridiculed, or quickly dismissed as unavoidable side effects. Yet a brief review of the food chains and biogeochemical cycles that maintain the balance of the biosphere should convince a thoughtful observer that the loss of plant species and wildlife can have far-reaching effects in addition to the loss of their aesthetic value.

The consequences of environmental contamination on the biosphere's organisms takes several forms: (1) direct kill; (2) indirect toxic effects resulting from biomagnification of substances in food chains or because of effects on reproductive performance or other metabolic systems or behavior; and (3) selective kill of certain species causing subsequent mortality at higher trophic levels through starvation or loss of shelter. The documented cases of environmental contamination are many; several books and many articles have been devoted solely to this subject. The incidents discussed in this chapter are only examples; additional references should be consulted for a more detailed review.

As described in the discussion of biogeochemical cycles in Chapter 2, materials deposited in one area are often rapidly circulated throughout the biosphere. So it is with pesticides. They travel via wind (drift), water (co-distillation and runoff), and in the bodies of migrating animals to sites far from their orginal application sites (see Figures 2-4 and 5-5).

Studies have shown that even under ideal conditions less than 50% of aerial sprays released at treetop level reach the target area (see Figure 7-24); the remaining portion is carried off in the wind, often many miles

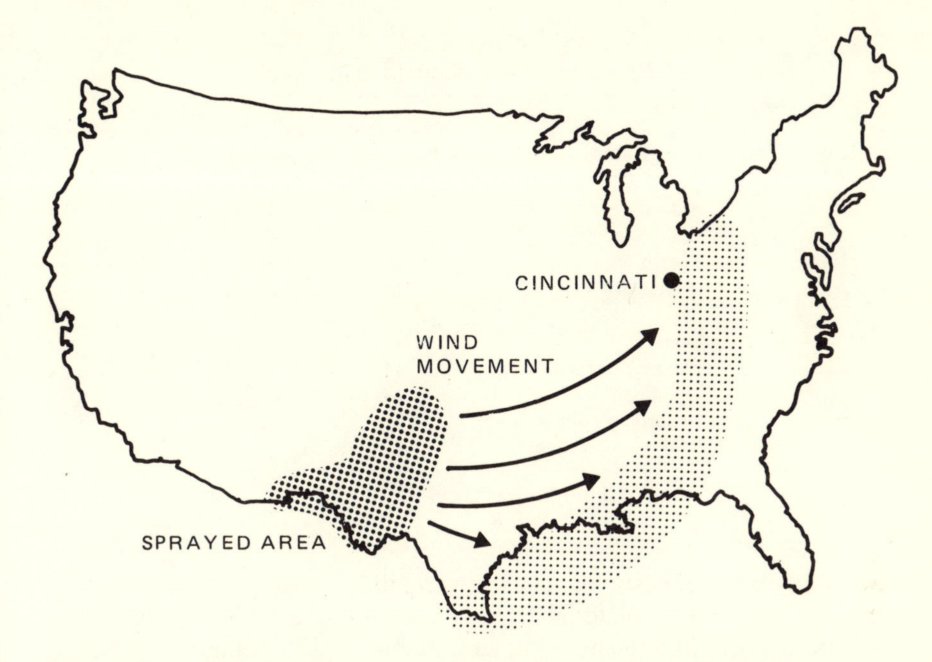

FIGURE 5-5. Long-distance transport of pesticides. Insecticides (DDT, DDE, chlordane, dieldrin, and heptachlor epoxide) sprayed in western Texas for agricultural pest control were picked up by a large dust storm in Texas and deposited in Cincinnati, Ohio, with a light rain the following day at a level of 1.3 ppm. At the time the dust fall occurred in Cincinnati, the dust cloud resulting from the storm stretched for 1500 miles in a 200-mile-wide band extending easterly from the southern tip of Texas north to Lake Erie (from Ehrlich and Ehrlich, 1970).

away. Much of that hitting the ground in the target area will be carried off in irrigation and rainwater or in the bodies of insects, birds, rodents, and other animals that migrate in and out of the application site. Thus it is clear that the environmental contamination caused by a pesticide application cannot be contained within a target area. For instance, significant quantities of persistent insecticides have been found in the bodies of Antarctic penguins inhabiting an area where no insecticide application has taken place for thousands of miles around.

Some nontarget organisms are killed immediately and directly by the pesticides. These include native plants and phytoplankton killed by herbicide application, drift, and runoff. Frequent direct-kill victims of insecticide application are beneficial insect species like bees and natural enemies of pests, other invertebrates, and small vertebrates such as birds and rodents resting on sprayed surfaces or feeding on contaminated vegetation.

TABLE 5-4
Persistence of Insecticides in Soils[a]

Insecticide	Years since treatment	Percent remaining
Aldrin	14	40
Chlordane	14	40
Endrin	14	41
Heptachlor	14	16
Dilan	14	23
Isodrin	14	15
Benzene hexachloride	14	10
Toxaphene	14	45
Dieldrin	15	31
DDT	17	39

[a] From Nash and Woolson (1967).

Many kinds of pesticides cause direct kill on nontarget species. Biomagnification of a toxic material in food chains is characteristic of pesticides that are persistent in the environment, mobile, lipid-soluble, and not rapidly degraded to less toxic substances. The best known of the persistent pesticides are the organochlorine insecticides like DDT, BHC, lindane, chlordane, heptachlor, aldrin–dieldrin, endrin, and toxaphene (see Table 5-4). Some cases of biomagnification in the food chain have become classics of environmental pollution, such as the buildup of DDT in the Lake Michigan ecosystem discussed briefly in Chapter 2 (Figure 2-11).

Another classic story of insecticide contamination and one of the earliest documented cases of biomagnification was observed at Clear Lake, California. Visitors to Clear Lake and residents of the area, popular holiday resort, were bothered by the presence of occasionally dense populations of a nonbiting gnat, the Clear Lake gnat. It was 1949 and the miracles of DDT were still being heralded; a popular notion at the time was that all bothersome insects could soon be eradicated by the new synthetic organic insecticides. Authorities examined the Clear Lake situation, chose a pesticide, DDD, which was not as directly toxic to fish as DDT, sprayed it on the lake, and scored what they felt was a spectacular success in the control of the Clear Lake gnat.

Two years later, in 1951, the gnat was back and the spray was repeated. Sprays continued at more frequent intervals until 1954, when the carcasses of large numbers of Western Grebes (a species of diving bird) began to accumulate in the lake area. Reproduction by the surviving grebes had been drastically curtailed.

How did it happen? The water in Clear Lake was found to contain

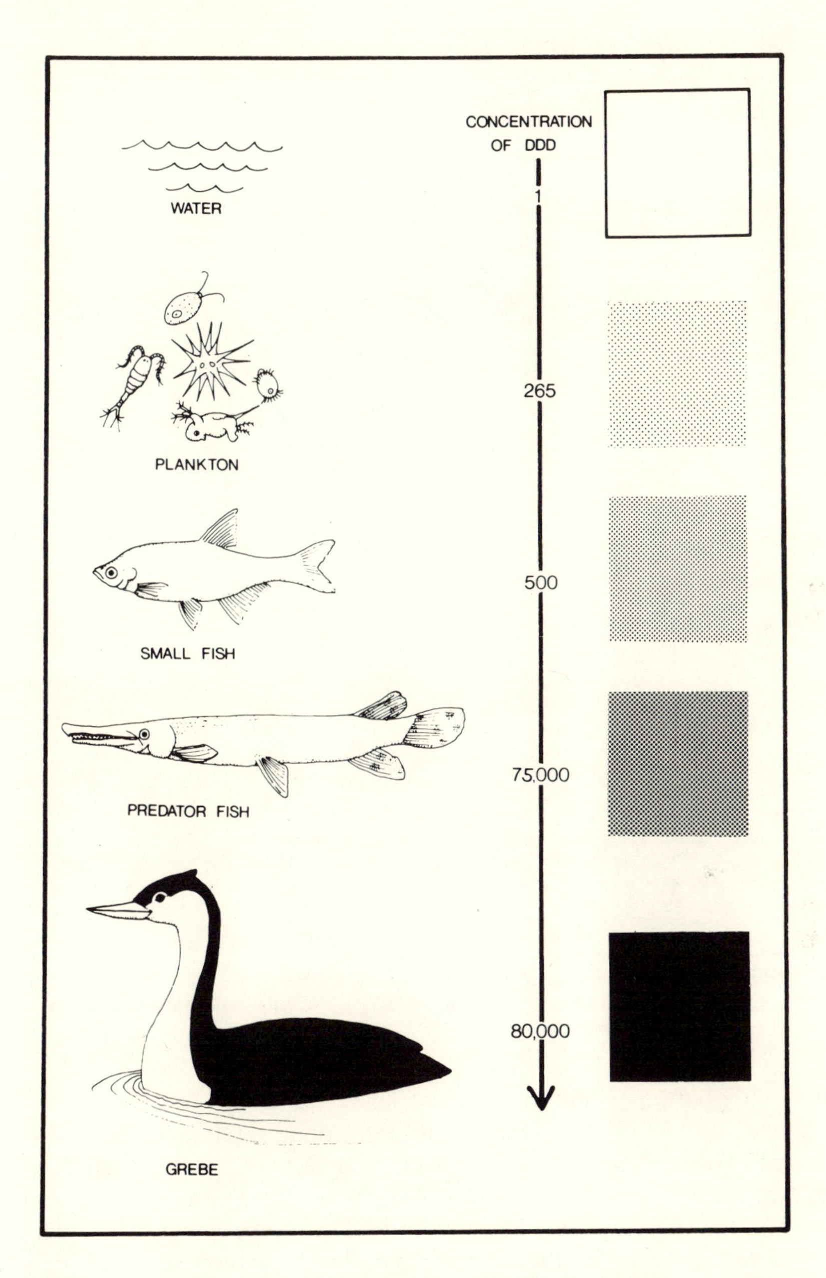

FIGURE 5-6. Accumulation of DDD in the Clear Lake food chain. DDD was concentrated in plankton at a level 265 times its concentration in water. In small fish the concentration was 500 times, in predator fish 75,000 times, and in grebes 80,000 times the DDD level in the lake's water.

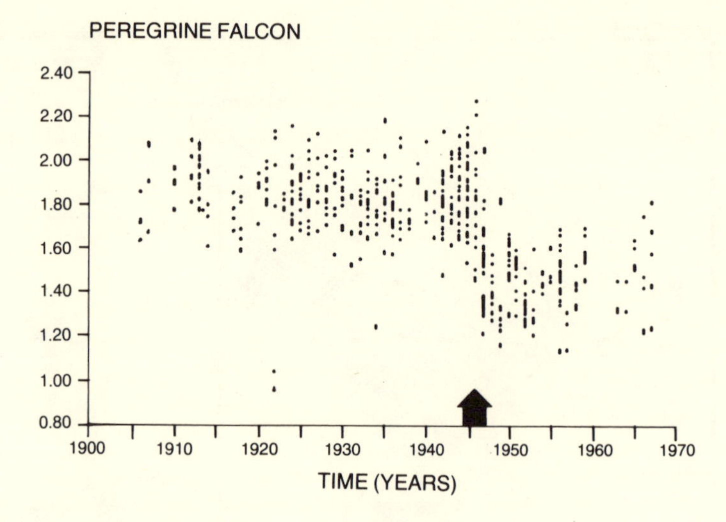

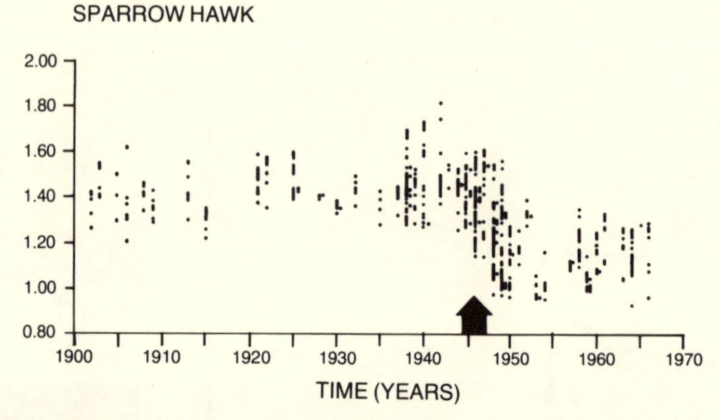

FIGURE 5-7. Changes in thickness of egg shells of the peregrine falcon and sparrow hawk in Great Britain. Arrows indicate first use of DDT (from Ehrlich and Ehrlich, 1970).

only 0.02 parts per million (ppm) of DDD, certainly not a directly toxic amount. Yet the plankton population had 5 ppm, small fish feeding on the plankton had several hundred times that amount of DDD in their tissues, and predatory fish were found to have up to 2000 ppm concentrated in their tissues (Figure 5-6). In other words, the grebes' diet of fish was so laced with poison that there could be no hope for their survival in Clear Lake for many years. The Clear Lake gnat story has a happy ending. In the late 1970s a fish was introduced into the lake to eat the gnat. Since that time, gnat populations have not been a problem and grebe numbers continue to increase.

Sometimes the poisoning resulting from accumulation of persistent pesticides takes a less obvious form—the inhibition of reproduction. Examples of this type of poisoning have been particularly well-documented in bird species: the peregrine falcon, the sparrow hawk, the brown pelican, the golden eagle, the bald eagle, and the osprey, to name a few. The numbers of offspring among all these species were declining after the 1940s, and this was in large part due to inhibition of reproduction. One of the more noticeable forms this inhibition takes is in the abnormally high breakage of eggshells before hatching time. Researchers have shown that this fragility was a result of decreased eggshell thickness (Figure 5-7) and lowered calcium content, and could be attributed to DDT's blockage of calcium metabolism in the mother bird. Since the banning of DDT in many parts of the western world, populations of some of the affected bird species have been increasing.

Not only are pesticides destructive to the biosphere through the direct or indirect poisoning of organisms, they may also affect the characteristic abundance of species by simplifying food webs or by destroying the food sources of many animals at higher trophic levels. Take the case of herbicides designed to kill broad-leaved species and thereby favor grains, grasses, conifers, and other narrow-leaved plants. All animals depend on plants for food at the very base of their food chain, and when any large portion of that energy source is removed, even for a short period of time, animals at all trophic levels suffer. Many animals can survive on only a few species of plants, and when these are destroyed to let other species move in, the whole food web of associated animals and microorganisms is disrupted.

The kill-off of insects and other animals by pesticides effects food webs in a similar manner. This has been well-documented in the case of natural enemies whose food source (the pest) has been abruptly reduced to nonsupportive numbers by pesticide application. The surviving natural enemies starve to death, emigrate, or quit reproducing, and the pest prospers, multiplying rapidly in a natural-enemy-free environment (see discussion of pest resurgence, Chapter 4).

HAZARD TO HEALTH

As in the case of environmental contamination it is difficult to assign a value to the impaired health and loss of human life resulting from pesticide poisoning. Partly because of this difficulty, this side effect of pesticide use has not been adequately considered in pest control decisions. It is often said that most cases of pesticide poisoning could have been avoided if label instructions were carefully followed; however, this contention too easily dismisses the fact that pesticide poisoning is common. Two health-related facts in the pesticide-use record are worth noting: (1) especially in the U.S. agricultural industry, many workers are either illiterate or can read only Spanish; and (2) accidents occur even among the most educated. In 1977 in California alone, over 1500 persons were reported suffering from pesticide poisoning; several times that many cases of nonoccupational poisonings probably went unreported. As in the case of any hazardous industry, carelessness is common and accidents occur in direct proportion to worker exposure to dangerous situations. Thus is stands to reason that if people had less contact with highly toxic pesticides, fewer poisonings would result. Pesticide poisoning in the home, especially among small children, is also far too frequent. Accordingly, we can consider it in the best interest of human health to minimize pesticide use wherever possible.

Many of the organophosphate insecticides like parathion, mevinphos, monocrotophos, TEPP, and dichlorvos are among the most hazardous materials with which to work. Although the organochlorines like DDT stay in the environment for long periods of time, cause environmental havoc through their buildup in food chains, and are in many instances suspected of having carcinogenic, tetratogenic, or mutagenic properties, they are much less acutely hazardous to the applicator than are the organophosphates. Typically, organophosphates depend on a quick, heavy kill immediately following application. Because they generally break down rapidly in the environment, they must often be applied repeatedly in a season to obtain the desired population reduction.

The long-range effects of pesticides on humans may include an increase in birth defects and in the occurrence of cancer. The herbicide 2,4,5-T commonly contains a contaminant dioxin known to be teratogenic and fetocidal to at least some mammals. Arsenic compounds used as insecticides have been reported to be carcinogens as well as direct poisons in slightly larger quantities. Three other fungicides derived (like the birth-defect-inducing drug thalidomide) from the chemical phthalimide—captan, difolatan, and folpet—have been shown to produce birth deformities in certain laboratory animals. Dichlorvos, the common active ingredient in flea collars, is also a suspected mutagen. Insecticides such as DDT, aldrin–dieldrin, and other organochlorines, as well as a number of other commonly used pesticides, have been implicated as probable carcinogens. The nematicide DBCP has been associated with a number of cases of sterility among men working in manufacturing and formulating plants. Knowledge of chemical induction of cancer and birth defects is still very scant. It is unfortunate but likely that as our knowledge grows, more pesticide materials will be implicated in these malignant disorders.

The need for an ecosystem outlook when carrying out pest control actions designed to protect human health can be illustrated by an incident which occurred in San Joaquin, Bolivia, in 1963. At this time more than 300 residents of this small town died of a viral disease called Bolivian hermorrhagic fever or "black typhus." An epidemic of this disease had never been known in the area before. The reservoir of the disease-causing virus was discovered to be a mouselike rodent called a laucha, large numbers of which had only recently become established in San Joaquin homes.

Why were these wild rodents suddenly able to move into town? It turned out that over the preceding five years the town's cat population, which had numbered in the several hundreds, had been mysteriously reduced to less than a dozen. Thus the rodents, released from the control of their natural enemies (the cats), were able to move into town for the first time.

But why had all the town's cats suddenly died? The answer lay in the impact of DDT on the cat population. As part of a malaria eradication program, the inside walls of all the houses had been sprayed with DDT to kill mosquitoes (transmitters of malaria-causing protozoa) that might enter the houses and land on the walls. However, the local cats picked up the DDT on their fur and then ingested it when grooming themselves. Eventually they accumulated lethal doses of the insecticide in their bodies and gradually almost all died. Thus, in San Joaquin, Bolivia, a pest control program designed to eliminate one serious disease—malaria—caused an epidemic of another—"black typhus."

The careful use of pesticides, of course, has resulted in many benefits to humanity over the last forty years through the control of disease, through the increased production of food and fiber, and as a result of the enhanced aesthetic quality of our recreational areas. Yet only with the conscientious integration of other control methods along with minimum use of chemical controls can we expect the benefits to outweigh the burdensome cost of pesticide use to our health and economic well-being, and to the environment.

CHAPTER 6

The Philosophy of Integrated Pest Management

The Strategy of the Future

While the ability of modern pesticides to annihilate pests locally is undeniable, the prevailing dependence on these pest control chemicals has repeatedly led to crisis situations (including pest resurgence, secondary pest outbreak, resistance, environmental contamination, and hazards to human health) that prove far worse than the original pest problem. In the last decade many have recognized the need for a new management approach that would minimize such negative side effects of pest control actions yet give effective, economical control of pest organisms. The new control strategy that has subsequently developed is Integrated Pest Management.

Integrated pest management is a popular, variously defined term. Three definitions will suffice to describe the approach taken in this text:

1. The U.N. Food and Agriculture Organization's (FAO) Panel of Experts on Integrated Pest Control (1967) defined IPM as:

> a pest management system that, in the context of the associated environment and the population dynamics of the pest species, utilizes all suitable techniques and methods in as compatible a manner as possible and maintains the pest populations at levels below those causing economic injury.

2. The Council on Environmental Quality (CEQ) in its publication *Integrated Pest Management* (1972) defined IPM as:

> an approach that employs a combination of techniques to control the wide variety of potential pests that may threaten crops. It involves maximum re-

liance on natural pest population controls, along with a combination of techniques that may contribute to suppression—cultural methods, pest-specific diseases, resistant crop varieties, sterile insects, attractants, augmentation of parasites or predators, or chemical pesticides as needed.

3. The definition given earlier in this book stated:

Integrated pest management (IPM) is an ecologically based pest control strategy that relies heavily on natural mortality factors such as natural enemies and weather and seeks out control tactics that disrupt these factors as little as possible. IPM uses pesticides, but only after systematic monitoring of pest populations and natural control factors indicates a need. Ideally, an integrated pest management program considers all available pest control actions, including no action, and evaluates the potential interaction among various control tactics, cultural practices, weather, other pests, and the crop to be protected.

The term *integrated pest management* as defined above is synonymous with the term "integrated control" as defined by Stern *et al.* (1959). However, integrated pest management should not be confused with the terms *pest management* or *insect pest management*. These terms refer to any practice or combination of practices designed to manipulate pest or potential pest populations and to diminish pest injury or render pests harmless. Pest management may involve simple manipulations such as spraying a rose bush with a pesticide or emptying water-filled tin cans to prevent mosquito breeding, or it may be accomplished through a complicated integrated pest management system.

Integrated pest management is a complex concept; this probably explains the existance of a variety of definitions—each emphasizing a different aspect of the strategy. The three definitions presented previously suggest several important aspects of the philosophy behind integrated pest management systems:

1. *A conception of the managed resource as a component of a functioning ecosystem.* Actions are taken to restore, preserve, or augment checks and balances in the system, not eliminate species. Surveys must be made to evaluate and avoid or diminish disruption of already existing natural controls of both the target pest and other potential pests. IPM programs do not include eradication methods, although it is recognized that in a very few instances eradication and not integrated pest management may be the best management strategy.
2. *An understanding that the presence of an organism of pestiferous capacity does not necessarily constitute a pest problem.* It must be ascertained, before a potentially disruptive control method is employed, that a pest problem actually exists. This requires the implementation of economic injury levels or some other suitable decision-making criterion.

3. *An automatic consideration of all possible pest control options before any action is taken.* The integrated pest management strategy utilizes a combination of all suitable techniques in as compatible a manner as possible; in other words, it is important that one control technique not antagonize another.

As living systems, managed ecosystems are dynamic, never static; the IPM concept recognizes this, and integrated pest management systems reflect this dynamism in their adaptability. Like any ecosystem, managed ecosystems are composed of a variety of biotic and abiotic components, some or many of which may be affected by any pest control action. Consequently, integrated pest management systems consider the whole ecosystem when making pest management decisions, not just the pest organism. Each managed situation is unique; that is, no two cotton fields, no two forests, or no two mosquito-breeding situations are identical. Each has a somewhat different community (biotic) composition and is subject to varying physical (abiotic) stresses. These differences between otherwise similar managed ecosystems may influence the eventual success of any pest control action. Integrated pest management systems try to evaluate these differences by surveying, sampling, and assessing each situation individually.

The key factor in the success of integrated pest management is not its tools or techniques, although these are extremely important, but the human element of the system—the decision-maker. Lack of well-trained pest managers is one of the major deterrents to the widescale use of integrated pest management. After years of ill-advised use of the preventive chemical control strategy—i.e., pesticides applied as dictated by a predetermined schedule—many pest managers are uncertain of their ability to make control decisions or are reluctant to spend the time and effort at regular intervals to survey the state of the resource ecosystem under their charge before taking pest control action. Yet a sound decision-making capacity is within the grasp of most resource managers with proper training and by following certain steps in setting up their individual integrated pest management programs.

A GUIDE FOR SETTING UP AN INTEGRATED PEST MANAGEMENT PROGRAM

1. *Understand the biology of the crop or resource, especially in the context of how it is influenced by the surrounding ecosystem.* This is very important, especially when later evaluating how and when significant damage to the resource may occur. Important considerations include the plant's type of life cycle: is it an annual, biennial, perennial? What initiates

growth at the beginning of the season—temperature, moisture, photoperiod, or a combination of the three? What extremes of temperature or moisture availability can kill the plant or cause it to go into a dormant state? If the resource involves the production of flowers, fruit, or seed, what requisites are necessary for instigation of flowering (e.g., photoperiod, temperature) and for setting of seed (e.g., pollinators such as bees or flies).

How does the crop plant respond to stress such as drought, nutrient deficiencies, and temperature? The resource manager should be able to recognize these stress signals and know how much stress the plant can tolerate. For example, some shade trees can be completely defoliated one season without significant loss of aesthetic value in the next, yet most annuals cannot withstand heavy or total defoliation.

What is the growth rate of the resource under various environmental conditions, and what is its seasonal growth cycle? For instance, many plants begin the season by putting most of their available energy into vegetative growth (roots, stalk, and leaves); then after flowering the energy sink switches its input to the fruiting parts of the plant (Figure 6-1). Certain conditions can retard or hasten fruiting and increase or decrease vegetative growth (Figure 6-2). It is important to be cognizant of these factors whether the crop is grown for foliage, stem-branch, root, or fruit production.

If plants are grown for their fruit, there is usually an optimum number of fruits each plant can produce. In some crops, if the number of fruit-producing buds can be limited, the plant will produce fewer but larger and, under current market grading systems, more marketable fruits.

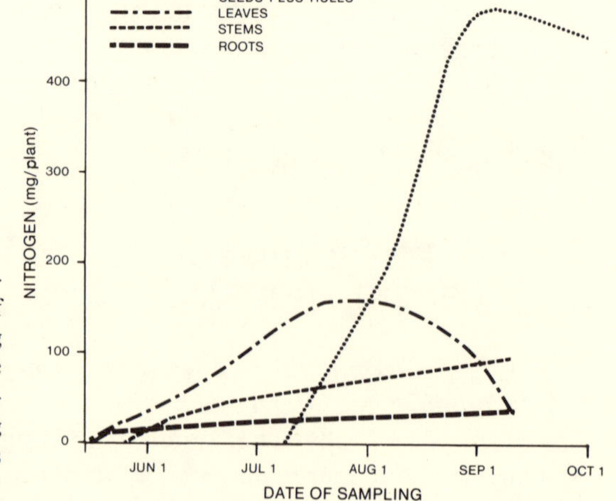

FIGURE 6-1. Changes in nitrogen content of various organs of the broad bean, *Vicia faba*, during growth. Later in the season the largest portion of nitrogen is transported to the fruits with a resulting loss of nitrogen from the leaves (after Salisbury and Ross, 1969).

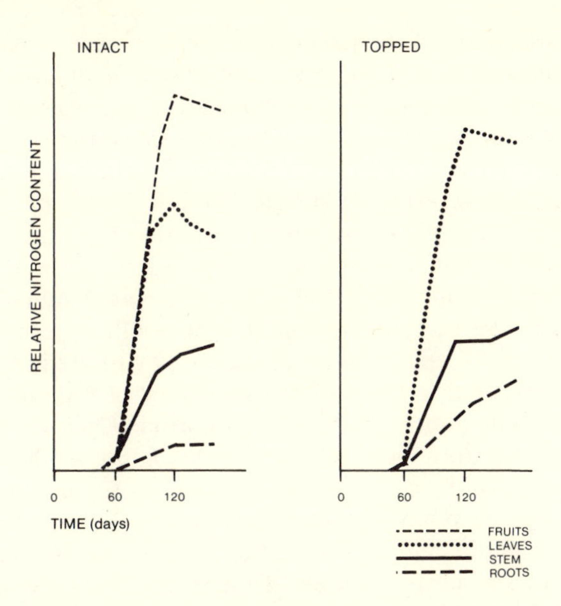

FIGURE 6-2. Topping of tobacco plants removes the developing fruits and reduces translocation of amino acids and other nitrogenous compounds from the marketable tobacco leaves.

Finally, it is important to know how the physical environment—e.g., soil type, nutrients, drainage, shading, weather—and various management practices—e.g., cultivation; irrigation; pruning; application of defoliants, fruiting inhibitors, and other toxic chemicals—affect the crop biology in this managed ecosystem.

2. *Identify the key pests; know their biology; recognize the kind of damage they inflict; and initiate studies on their economic status.* Key pests, as defined in Chapter 3, are organisms that cause a significant reduction in resource yield or quality every season unless some pest management action is taken to control them. These are the pests around which integrated pest management systems are built. Usually there are only one or two key insect pests in a managed resource system; other pests may cause sporadic damage.

Key pests are not always the most numerous species in the managed resource ecosystem but are the ones that cause the most significant damage most of the time. For instance, an aphid species feeding on leaf sap may be far more numerous in an apple orchard than a caterpillar that bores into the fruit, yet it may be that the caterpillar, causing direct damage to marketable fruit, is a key pest and the aphid virtually innocuous. In other words, the classification of a species as a key pest is dependent on the synchronization of its damaging stage (e.g., caterpillar) with the vulnerable stage of the resource (e.g., fruit), the type of damage the pest causes, the plant's tolerance of that damage, the consumer's tolerance of that damage, and the damage potential of a single pest individual.

Key pests are rarely, if ever, universal to given resource ecosystems. Their presence and severity of damage is often limited by climate and other local ecosystem variables. An agricultural crop like cotton illustrates this clearly (Figure 6-3). In the humid regions of the southeastern U.S. the boll weevil and a plant bug (*Lygus lineolaris*) are considered to be key pests; in the semiarid regions of the southwestern U.S. (e.g., Texas) the boll weevil and fleahopper are the key pests; in the irrigated deserts of the far west (e.g., the Imperial Valley of California) the key pest is the pink bollworm; and a bit further north in California's southern San Joaquin Valley only the lygus bug (*Lygus hesperus*) has been considered a key pest, in some situations erroneously so. In reality, the intensity of damage from any of these pests may vary from locality to locality and from field to field within each of these areas. In fact, even with the most severe key pests, it often happens that many fields escape a significantly damaging infestation. Thus it is critical that control action

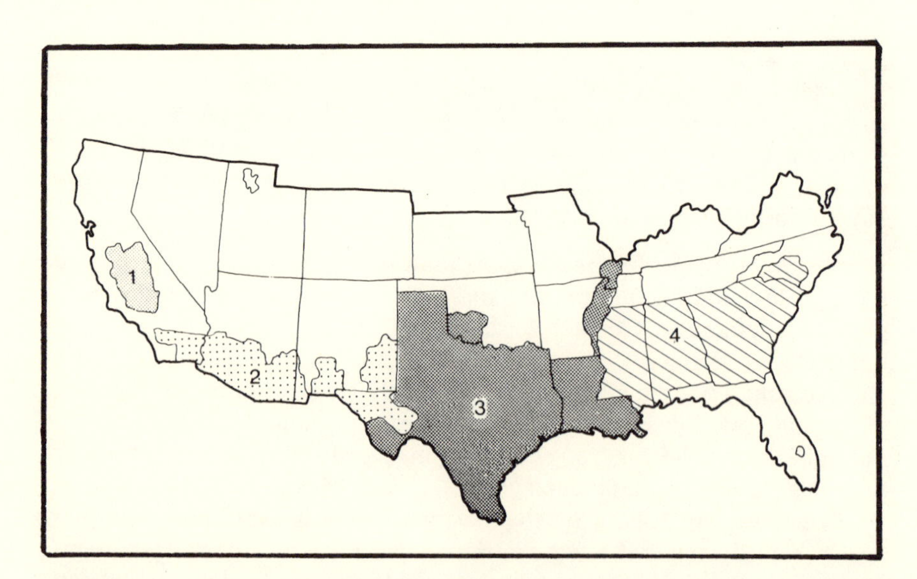

FIGURE 6-3. Cotton-growing areas of the U.S. showing key pests. Region 1: San Joaquin Valley of California: lygus bugs. Region 2: Irrigated deserts of the far West: pink bollworm. Region 3: Central Southwest: boll weevil, fleahopper. Region 4: Southeast: boll weevil, plant bug. [*Heliothis* spp. (the cotton bollworm and the tobacco budworm) were previously considered key pests in all these regions. However, *Heliothis* is apparently a secondary induced pest in all cotton-growing areas of the U.S.]

thresholds be established and constantly reevaluated within the managed ecosystem.

3. *Consider and identify as quickly as possible the key environmental factors that impinge (favorably or unfavorably) upon pest and potential pest species in the ecosystem.* What factors limit survival, development, and reproduction of the key pests in the given ecosystem? Major limiting factors usually include natural enemies (parasites, predators, and pathogens) of the pest, and, as has been noted in earlier chapters, these limiting factors often exert more intense pressure on the pest population as it increases in abundance. The availability of proper food sources may affect pest populations. For instance, the reproductive capacity of the soybean looper is greatly enhanced when cotton flowers are available to provide an extra source of food (nectar) for the adult female moths, and it has been shown that lack of a nectar source so diminishes the reproductive capacity of the loopers that soybeans grown at great distances from cotton fields rarely have serious infestations of this pest.

Temperature, water availability, photoperiod, shelter, and other abiotic factors may also limit pest populations. Particularly important in the survival of some insect pests is the availability of overwintering sites. For example, following harvest the pink bollworm usually overwinters as a last instar larva in cotton bolls missed in picking or within the cotton stalk. Destruction of these overwintering sites has been an important component in pink bollworm management in some areas of its range.

4. *Consider concepts, methods, and materials that individually and in combination will help to suppress permanently or restrain pest and potential pest species.* A pest's injuriousness can be permanently diminished by lowering its equilibrium position. Several ways of doing this were discussed in Chapter 3, including the introduction and establishment of a new natural enemy in the ecosystem or permanently altering the pest's physical environment so as to reduce its reproduction and survival (e.g., removal of breeding sites and refuges). The use of such permanent control methods are the most effective and, over a long period of time, the most economical tactics contributing to pest population management.

5. *Structure the program so that it will have the flexibility required to adjust to change, i.e., avoid a rigid program that cannot be adjusted to variations from field to field, area to area, or year to year.* No two pest situations are the same; often there will be significant differences in pest populations even between neighboring fields. The same field or orchard will vary from year to year in the character or severity of its pest problems. In other words, a pest control program that worked well last year might be entirely inappropriate for this year's pest complex. A good

integrated pest management system recognizes this variability and can be adjusted easily to such changes.

First, the operational program must detect these changes and variations. Detection is effected through the implementation of a good monitoring system to record and gauge plant growth, fluctuations in pest and natural enemy populations, the weather, and other pertinent variables. Second, the program must offer various control action options and combinations of options according to the severity and nature of the problem as revealed by the monitoring results. These might include cultural actions (e.g., early harvest), relying on a natural mortality factor, and selective or application of a pesticide.

6. *Anticipate unforeseen developments; expect setbacks; move with caution; above all, be constantly aware of the complexity of the resource ecosystem and the changes that can occur within it.* The integrated pest management specialist must keep a close tab on the pulse of the managed ecosystem. He/she must recognize the first signs of possible change—e.g., presence of eggs of a new herbivore, arrival of migrating predators, a change in plant character that indicates a new stress (drought, nutritional deficiency, pest damage). When sudden disruptions occur in the surrounding ecosystem—e.g., harvest or pesticide application in a neighboring field, unexpected rain or a freeze—the pest manager must be particularly on the lookout for trouble. And any time a control action or other management operation within his/her own managed system is taken, the pest manager must closely evaluate its effects. If a control action did not have the expected results (i.e., reduction of pest damage) this could indicate other troubles in the ecosystem (e.g., resistance of pest to pesticides, immigration of pest individuals, poor application technique).

7. *Seek the weak links in the life cycle of the key pest species and direct deliberate control practices as narrowly as possible at these weak links. Avoid broad impact on the resource ecosystem.* When is the pest species most vulnerable? Only a pest manager familiar with the biology of the pest in the local resource ecosystem can answer that question. The tobacco hornworm is most vulnerable right after tobacco harvest. At this point it must find enough food to complete its last larval stages before burrowing deep beneath the soil to overwinter as a pupa. If the tobacco farmer removes all tobacco plant debris immediately after harvest, he is taking advantage of the hornworm's vulnerability and will greatly decrease next year's problem. If all growers in the region cooperate in such crop debris destruction, even better control can be achieved.

Control for the Japanese beetle is most easily carried out during its larval or grub stage with applications of a commercially available pathogen "milky spore disease" applied to the beetle grub's sod habitat (Figure 6-4). As an adult this insect is winged and very mobile, flying from tree to tree and shrub to shrub, and consequently much more difficult to control; additionally, milky disease is ineffective against adult Japanese beetles.

Pesticides are often most effective at a certain stage in a pest's life cycle. Insecticides are sometimes most destructive of young larvae and in some cases do not give satisfactory control of older larvae, pupae, adults, or eggs; the difference may be especially accentuated if older larvae become inaccessible to surface-applied materials. Thus the susceptibility of the prevalent stage of the pest insect should be known before insecticide application is made.

Many pest species are readily managed by various forms of cultural control. Some insects, such as wireworms and the plum curculio, are vulnerable to the direct effects of tillage when they occur in the soil as prepupal or pupae. Crop rotation takes advantage of the narrow dietary requirements of many pest organisms and exploits this type of vulnerability; this cultural technique may be effective against some plant pathogens (e.g., the bean–barley rotation against the fusarium, rhizoctonia, and thielaviopsis root rots of bean) and nematode pests as well as insects—the corn rootworm being a good example of the latter. Removal of alternate hosts necessary for the winter or summer survival of a pest can also severely limit its survival. Use of sticky bands around tree trunks take advantage of vulnerable stages of insects that move up or down tree trunks—e.g., gypsy moth larvae, ants, and elm leaf beetle larvae.

Natural enemies of pests have evolved habits that exploit the vulnerability of their hosts. Careful management of naturally existing predators and parasites and the introduction of new natural enemy species enhance this exploitation of weaknesses in the pest's line of defense.

Another set of weaknesses in a pest insect's life cycle that has been effectively utilized in integrated pest management systems is its mode of sexual stimulation. Sex attractants are used frequently to trap (usually for monitoring purposes) particular species of pest insects.

8. *Whenever possible, consider and develop methods that preserve, complement, and augment the biotic and physical mortality factors that characterize the ecosystem.* This is an expansion of the previous point. Many cultural practices are designed to intensify further the hardships presented by the more brutal seasons of the year. For example, postharvest cultivation exposes overwintering larvae and pupae to predators, cold, heat, and desiccation; provision of nesting boxes encourages insect-consuming birds or wasps to forage in particular forest or field areas; supplementing food sources for parasites (e.g., by providing a year-round supply of nectar-producing flowers) or predators (e.g., by placing artificial

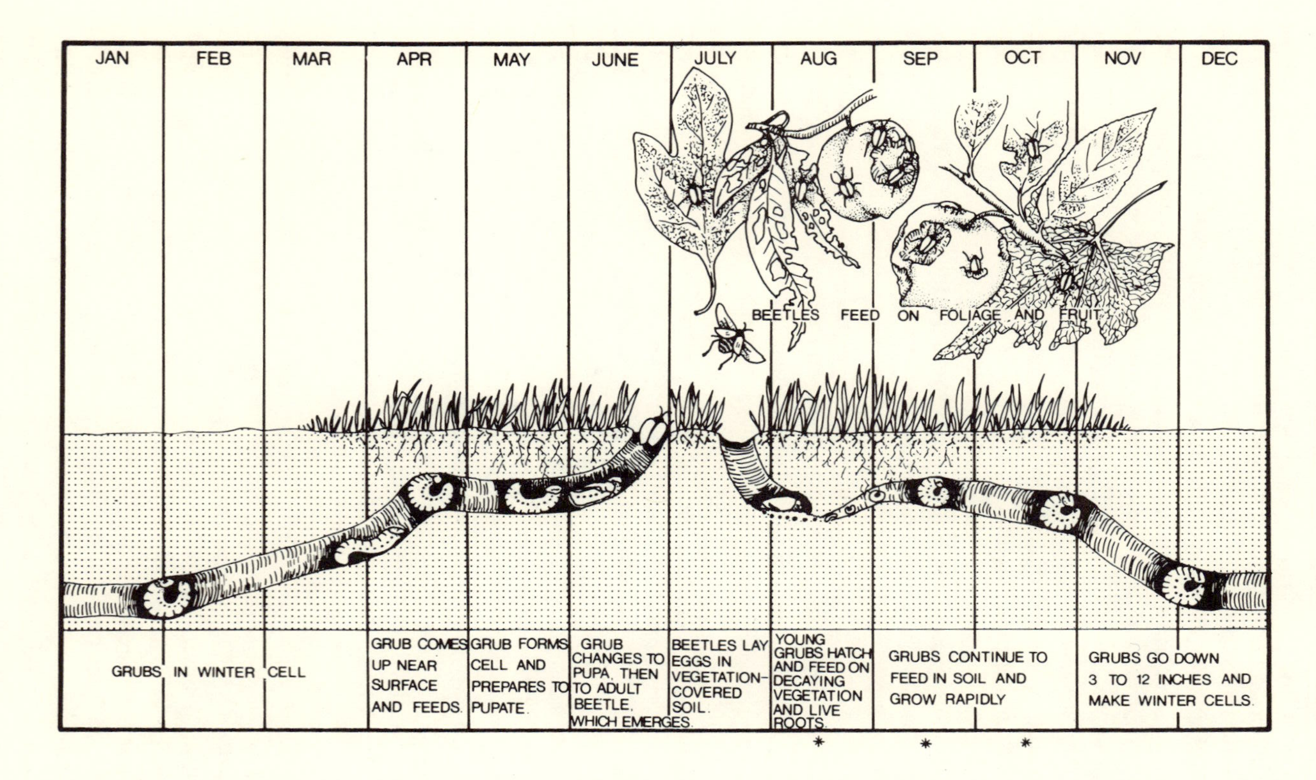

FIGURE 6-4. Life cycle of the Japanese beetle. The most effective and environmentally sound control can be obtained by treatment with "milky spore disease" at the young grub stage, when the grubs feed on sod and debris (starred months) (after Metcalf and Flint, 1951).

food in the fields) can greatly enhance the effectiveness of these natural mortality factors.

Strategies involving early or late planting of a crop often increase vulnerability of pests to various mortality factors. Planting an early or a postharvest trap crop may be used to prevent pests from entering the main crop or going into a protected state; they can be easily destroyed along with the trap crop. This has been a useful control method for overwintered insect pests of soybean (early trap) and for witchweed (postharvest trap), a plant parasitic on a diversity of crops such as tomato and sorghum.

Another important strategy that preserves naturally occurring mortality factors is the selective use of insecticides. Selective insecticides kill the target pest species and/or enemies of other pests or potential pests in the ecosystem. Proper timing or placement of insecticide applications (ecologically selective use of pesticides) can often achieve the same goal.

9. *Whenever feasible, attempt to diversify the ecosystem.* Compared to natural ecosystems, diversity has been decreased on all levels in managed ecosystems. Genetic and age diversity of the crop, species diversity of the community, and physical diversity of the environment have all been decreased to provide the most efficient system possible for maximizing the resource. In general, this is both good and necessary. However, with a decrease in diversity in an ecosystem, there is also usually, but not always, a decrease in that ecosystem's stability or its ability to resist new kinds of stress. This can be catastrophic in certain situations.

For instance, a variety of corn planted throughout the Midwest in the late 1960s was particularly susceptible to a bacterial disease, corn leaf blight. Consequently, an epidemic of the disease was able to spread rapidly over the Cornbelt with devastating effects on the crop. If corn grown in these areas had been more genetically diverse in its resistance to the disease, the epidemic would not have been so widespread and severe.

Diversifying the ecosystem can be important in encouraging the role of natural enemies. At times the simple addition of one alternate food source or a shelter area can make the difference between effective and ineffective biological control. Such a case was presented in Chapter 3 with the discussion of how, in the vineyard ecosystem, the simple addition of blackberry bushes, which provided an alternate host for overwintering grape leafhopper parasites, was able to enhance the control of the grape leafhopper. Similarly, in apple orchards in Michigan, researchers have discovered that a sod or weedy cover beneath the trees provides an overwintering environment favorable to a highly efficient predatory mite that moves in the spring into the trees and controls phytophagous mites. Similarly, a ground cover containing nectar-bearing flowers has benefited

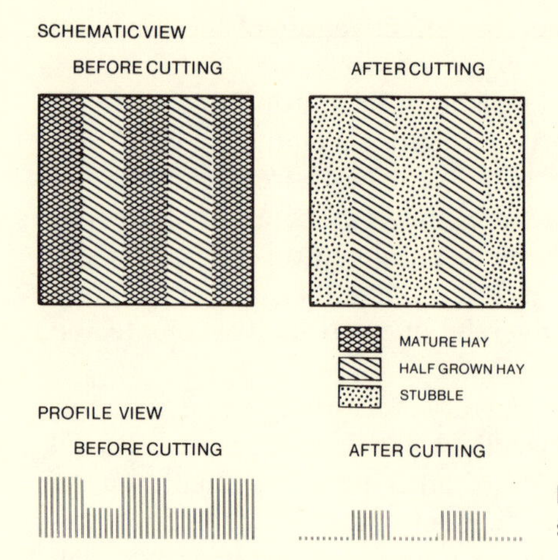

FIGURE 6-5. Schematic diagram of the system of strip harvesting alfalfa.

hymenopteran parasites of San Jose scale in deciduous fruit orchards in Europe.

In alfalfa, age structure of plants has been diversified and parasites and predators subsequently favored by the practice of strip harvesting adjacent rows of the crop (Figures 6-5 and 6-6). The uncut strips provide shelter and food for natural enemies that migrate, starve, or die of exposure in normally harvested or full-cut alfalfa fields. Such strip harvesting also prevents the mass migration of lygus bugs from alfalfa, where

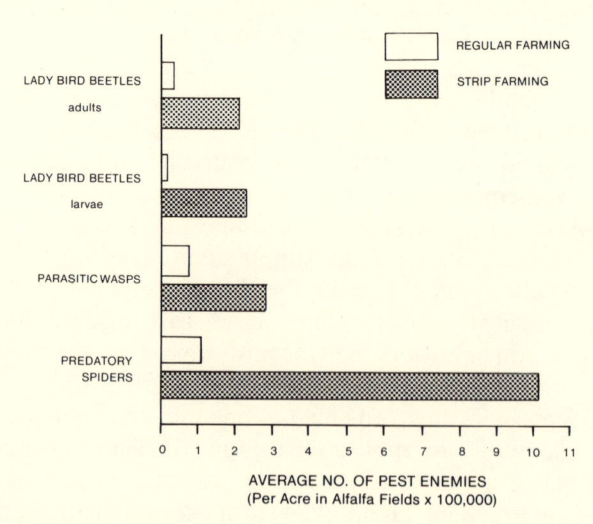

FIGURE 6-6. Effect of strip harvesting on natural enemies in the alfalfa field (from Council on Environmental Quality, 1972).

they do little damage to the hay crop, into adjacent fields of cotton where they may cause severe damage. Similarly, interplanting strips of alfalfa in cotton fields has been shown to draw lygus bugs out of the cotton to the alfalfa plants, their preferred food source.

It is particularly important, however, to proceed with caution when diversifying a managed ecosystem. The introduction of a new plant species may also provide alternate hosts for serious pests, especially plant pathogens, or result in new weed problems. These possibilities must be carefully explored by the integrated pest manager before taking such action.

10. *Assume and even insist that technical surveillance for programs be available (i.e., monitoring).* Effective monitoring is absolutely essential to the success of an integrated pest management program. There is no way to know what is going on in the managed ecosystem without regular, careful sampling of the pests and natural enemies, and assessment of crop performance in each area of the managed resource. Since even adjacent fields, forest areas, or city neighborhoods vary in soil type, moisture availability, and animal and plant inhabitants, sampling just one or two representative areas is not enough. Each must be individually surveyed and evaluated. This requires well-trained professional pest management consultants. Although it might be tempting to try to make economic savings in this area, it is always a mistake to monitor the resource inadequately. Poorly trained, overworked, or biased "advisors" may tend to take too few or less careful samples or miss important clues to future problems in individual portions of the managed ecosystem.

If the guidelines presented in this chapter are carefully followed in setting up and executing an integrated pest management program, the resource manager is on his way to effective, economical, and environmentally sound pest control. The next chapter describes tools and techniques appropriate for such programs.

CHAPTER 7

Practical Procedures

IPM Monitoring, Decision-Making, and the Tools and Techniques of the Integrated Pest Manager

In Chapter 1, it was stated that an integrated pest management program is composed of six basic elements: (1) the pest manager; (2) knowledge–information; (3) monitoring of the numbers and state of the ecosystem elements, e.g., weather, resource, pest natural enemies; (4) decision-making levels, the pest densities at which control methods are put into action; (5) IPM methods, the techniques used to manipulate pest populations; and (6) agents and materials, the tools of manipulation. The critical importance of the first two elements has been stressed in earlier chapters, and the need for a large corps of well-trained pest management specialists is further eleborated in Chapter 9. This chapter focuses on the remaining four elements of IPM: monitoring, decision-making levels and other decision-making guides, and the tools and techniques involved in manipulating pest populations to enhance production of the resource.

MONITORING AND SAMPLING; DATA COLLECTION: READING THE PULSE OF THE ECOSYSTEM

The pest manager must have an understanding of the biology and ecology of the crop or resource and its pests, and an awareness of how both are regulated or influenced by other factors in the same and surrounding ecosystems; this is the knowledge–information element of integrated pest management. In addition, the pest manager must also be constantly in touch with the changes occurring in the state of the resource or crop and in the numbers of pest and beneficial organisms present.

As earlier chapters have illustrated, each management situation is unique. Different fields differ in soil type, moisture availability, influences from surrounding ecosystems, management practices, and a host of other variables that can directly or indirectly influence past populations. Likewise, these influences change over time—often unpredictably—according to the situation. To keep in touch with these inconsistencies and fluctuations, the pest manager must monitor the managed ecosystem on a continuing basis.

Two kinds of monitoring programs are important in the development of integrated pest management systems: Monitoring for research purposes—e.g., to establish economic thresholds; to evaluate the influence of weather, natural enemies, or other natural suppression factors; to gauge the effectiveness of a potential control action; or to determine the most damaging stage of the pest—is exploratory and requires an extremely sensitive and careful watch on all the ecosystem components. On the other hand, monitoring systems for use by pest managers in operational, commercial IPM situations must be as quick, inexpensive, and simple in execution as possible, while still giving an evaluation of the field situation that is both accurate and useful. This chapter discusses sampling methods that can be useful to the pest manager in the actual implementation of an IPM program; sampling procedures designed for research purposes are reviewed by Southwood (1966) and Ruesink and Kogan (1975), and in several other sources. The sampling methods discussed below have been designed primarily for arthropod pests. Systems for monitoring other pests (with a few exceptions, such as for fire blight of pears and apple scab) have not been well worked out and, for the most part, have not been incorporated into working control programs.

Currently, the four most useful ways to sample for pest insects are random sampling, point sampling, trap sampling, and sequential sampling. The choice of sampling method depends on the crop and pest, the accuracy required, and the labor available. In some cases, a combination of two may prove useful, especially when more than one pest is involved. In other cases, an entirely new system may have to be evolved to deal with peculiar problems in the pest–resource situation.

The *random sample* is the most commonly used monitoring design in current integrated pest management programs. It can be used to determine pest numbers and/or damage per sample unit. Counts of pest numbers, damage, and/or natural enemy populations are taken at random spots in the field, forest section, or other managed unit. Usually about

4 to 6 or more spots are chosen randomly for sampling in a 40- to 80-acre field. Spots should be chosen so they are not too close to the edge of the field, and if samples are taken from irregular areas in the field such as particularly lush spots, these irregularities should be noted along with the sampling results. Even while carrying out a random sample program, field checkers should be on the lookout for unusual problems or conditions in the rest of the field. Figure 7-1 shows an example of a random sample plan. While different sampling spots should be chosen each time the field is surveyed (e.g., weekly), field workers should continue to sample "hot spots" with concentrations of pests until natural enemies or pest control activity reduces the population.

A second type of sampling is *point sampling*. In this type of sampling, pests and natural enemies are counted along with a more detailed monitoring of the crop maturity state in one or more areas of the field. For instance, the pest manager might choose (at random again) three rows of plants in a cotton field. Then he would examine the first 50 squares (fruiting buds) along each row, checking them for numbers of bollworms, bollworm eggs, or natural enemies, and also noting the percentage of damaged squares among those first 50 encountered. An estimation of the frequency

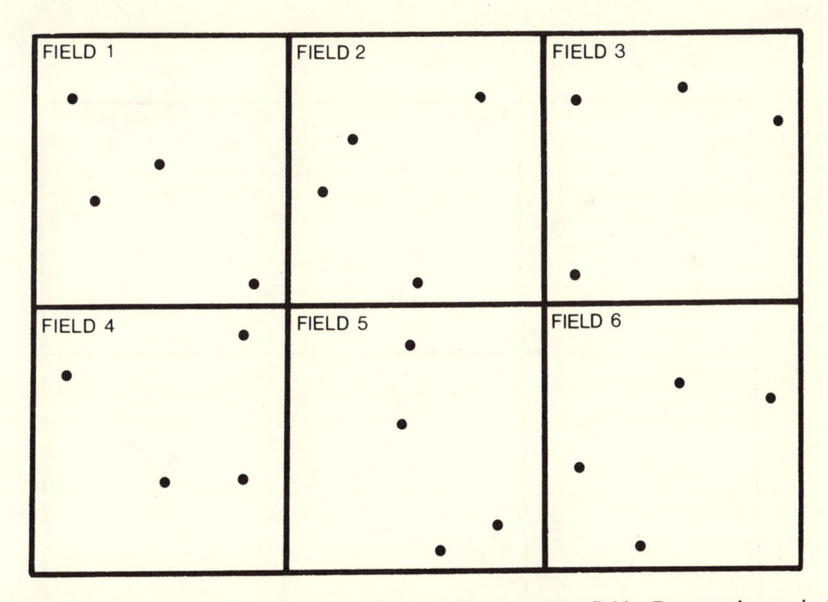

FIGURE 7-1. Random sampling in six adjacent 40-acre cotton fields. Dots mark a typical scheme of sampling spots that might be chosen by a pest manager for a routine sampling. At each spot the sampler would examine 25 cotton plant terminals for the presence of bollworm, bollworm eggs, or natural enemies. The next time monitoring takes place, four new spots, chosen at random, would be sampled.

of these fruiting buds is made by measuring how many feet of the row must be examined before 50 are counted. As the crop matures, producing more squares each week, the row length of the sampling sites shrinks, and the pest manager is able to estimate the maturity of the plants. A knowledge of crop maturity is often important, as removal of fruiting parts by pests may prolong the vegetative stage of the plant and delay harvest. Also, in many crops, after a certain point of maturity, some pests can no longer inflict economic damage at any population density (e.g., the tobacco budworm in tobacco; see Chapter 3). Figure 7-2 shows a point sampling scheme in cotton.

A third way to sample for pest populations is with the aid of *traps*. These are most commonly employed for detecting the actual presence of pest species and at present are not as useful as other methods for estimating pest densities. One of their most valuable uses is to detect the first appearance of migrating insects or to keep a check on the progress of the pest's life cycle—especially in determining the emergence dates of adult egg-laying stages. A common type of trap used for this purpose is the *light trap* (Figure 7-3A), which attracts nocturnal flying insects such as many moths and mosquitoes. With such traps, insects are attracted to an ultraviolet fluorescent light and fall into a collection jar containing a killing agent.

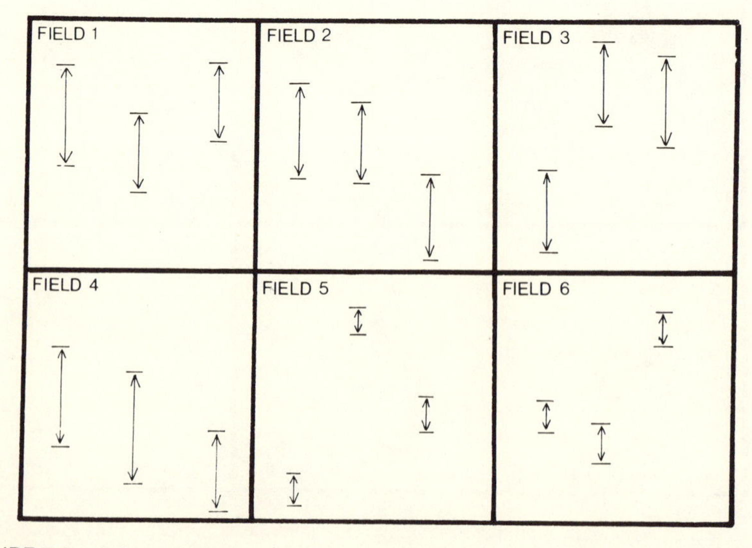

FIGURE 7-2. Point sampling along rows of cotton plants in six adjacent cotton fields. Each row is sampled for pests, damage, and natural enemies until 50 fruiting buds are encountered. The shorter length of sampled rows in fields 5 and 6 indicates that these plants have more fruiting buds than the other fields and are thus closer to maturity.

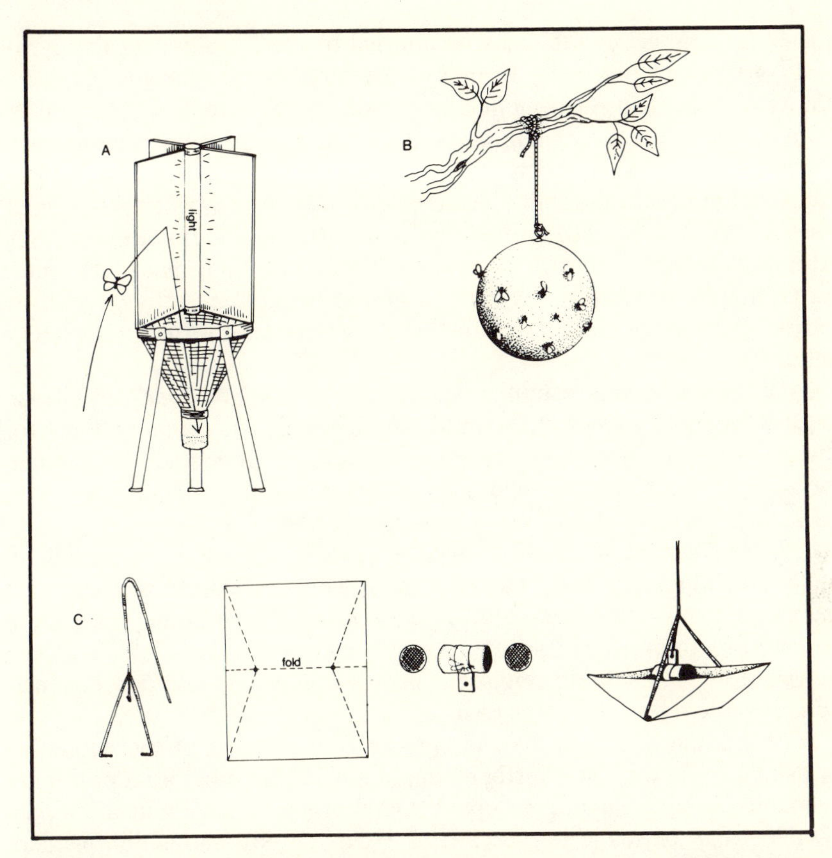

FIGURE 7-3. Useful traps for sampling and detecting pest insects. (A) A light trap. (B) A sticky red sphere. (C) A pheromone trap, showing how it can be made with wire and cardboard and baited with a virgin female placed in a small screened cage in the center.

Sticky traps have been used for many species of pests. Fly paper is probably the most popularly known of these sticky traps. Sticky red spheres (Figure 7-3B) have been used to estimate adult apple maggot abundance in orchards in Michigan and have also been successfully employed as a control for these pests in small orchards in New England. The maggot is apparently attracted by the red sphere's resemblance to a ripe apple.

Among the various kinds of traps, those utilizing *pheromone attractants* probably hold the most potential for widespread use in future integrated pest management programs. These traps exploit the chemical olfactory stimulants (pheromones) insects use in communication. Sex-attractant pheromones have been the most useful to date. Such sex pheromones may be excreted by the male (e.g., the boll weevil) and used to attract the female, or they may be emitted by the female (e.g., the gypsy moth) to attract the male. In many cases, chemical analogues to these naturally produced pheromones have been synthesized; in other cases, traps are baited with virgin females (excreting attractive pheromones naturally) in screened cages. Insects flying into the traps get stuck on a sticky substance on the bottom or sides of the trap. Figure 7-3C shows a pheromone trap, baited with a live female, used for codling moth detection. Although accurate density correlations have not been made with such traps, they have proven extremely useful in pinpointing activity peaks in various pest species and thus allowing the advantageous timing of control actions.

Other attractants useful in trapping pests include *food lures* (e.g., methyl eugenol for the male oriental fruit fly, sugar and propiononitrile for the house fly, coumarin for the sweet clover weevil, sinigrin for the diamond back moth, protein hydrolysate for female fruit flies). Traps may be baited with actual food or ovipositional lures as well.

A fourth method of sampling is *sequential sampling*. Although sequential sampling systems have not been well worked out in many pest management systems at present, they will probably become much more common in the future. Especially when pest populations are very high or very low, sequential sampling can save much time and consequently considerable labor and labor costs.

Sequential sampling systems utilize a knowledge of the economic threshold to determine if further sampling is required before a pest management decision should be made. Quick decisions without additional sampling can be made in the case of extremely low (no action needed) or extremely high (pest control action required) pest densities, but more time and perhaps more sampling will be required to properly evaluate moderate pest populations. For instance, Figure 7-4 illustrates sequential sampling systems for the green cloverworm in three growth stages of soybeans. At the first growth stage (i.e., 7 nodes present) the economic threshold is 12 larvae/foot of row. Thus, if after sampling only a few row-feet the pest manager finds populations of cloverworm larvae well below this number (e.g., 0–6/foot-row), he can go on to the next field confident that no control action is needed at this time. If numbers are well above this threshold, the need for a control action is certain and sampling can be stopped. If numbers of larvae are close to the threshold, more samples are required to properly evaluate the situation. Sequential sampling models can be translated into tables. Table 7-1 is a tabular representation of the sequential sampling model shown in Figure 7-5 for a general predator in soybeans (*Nabis* spp.). This tabular form of sequential sampling values is much more convenient for the pest manager to use in the field.

Many sampling techniques have been used in conjunction with ran-

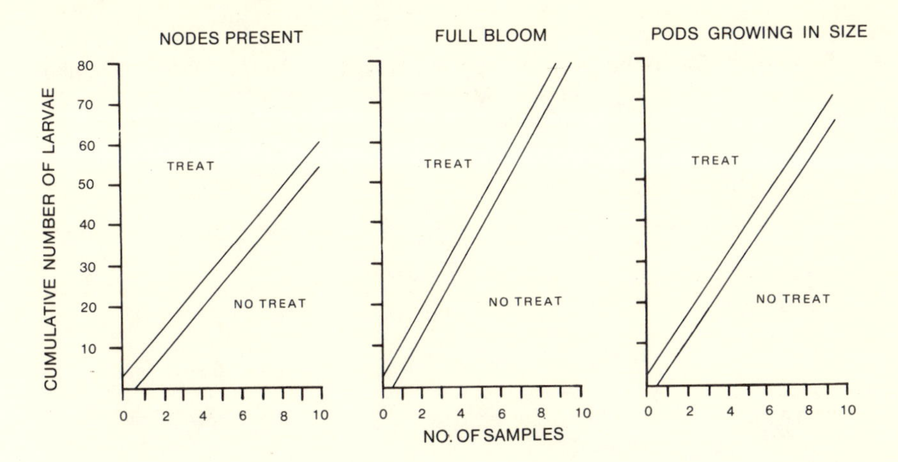

FIGURE 7-4. Sequential sampling plans for green cloverworm larvae at three soybean plant growth stages. Horizontal axis shows number of samples taken, vertical axis shows cumulative number of larvae. Sampling is continued until number of larvae collected falls above or below diagonal lines (i.e., in TREAT or NO TREAT sectors) (from Hammond and Pedigo, 1976).

dom, point, and sequential sampling schemes. The *visual count* method is the most commonly used. Visual methods include counting numbers of pest per leaf, per terminal, per square foot of trunk bark, or per similar plant unit. Visual methods are also used to assess damage per plant part or unit area.

The use of the *sweep net* is another common technique for sampling insect populations, especially for highly mobile insects such as plant bugs, that might jump off the plant before a visual count can be made. A major problem in using the sweep net is that of standardizing the sweep—particularly if more than one individual is doing the sampling. Another problem is that insects move up and down within plants during different times of the day, so sampling times must be consistant. But this is true of other sampling techniques as well. Ruesink and Kogan (1975) give four clues for standardizing sweeping technique in some crops:

1. Use a pendulum swing, as if you were sweeping the sidewalk with a broom.
2. In short vegetation, swing the net as deep as possible without taking too much dirt into the net. In tall vegatation, sweep only deep enough to keep the upper edge of the sweep net opening even with the top of the vegetation.
3. Sweep one stroke per step while walking at a casual pace.
4. Use a sweep net having a 15-inch diameter opening.

Table 7-1
Sequential Sampling of *Nabis* spp. on Soybeans by the Modified Shake Method[a,b]

Number of 4-foot samples[c]		Cumulative number of *Nabis* spp.							
		≤		≥		≤		≥	
1		5		—		—		29	
2		13		19		21		45	
3		21		26		37		62	
4		29		34		54		78	
5		37		42		70		94	
6		45		50		87		111	
7	LOW	53	CONTINUE SAMPLING	58	MEDIUM	103	CONTINUE SAMPLING	127	HIGH
8		61		66		120		143	
9		69		74		136		160	
10		77		82		153		177	
11		85		90		169		193	
12		93		98		186		210	
13		101		106		202		226	
14		109		114		219		243	
15		117		122		235		259	

[a] From Waddill *et al.* (1974).
[b] As *Nabis* is a valuable natural enemy in the soybean field, this system evaluates biological control potential for the *Nabis* population (as "high," "medium," or "low") rather than making a "treat" or "no treat" decision. This information, combined with pest sampling results, can aid in making a pest control decision.
[c] A sample is defined as one 4-foot section of row beat over a ground-cloth.

Various crops may require different sweep net techniques. For example, in cotton, the net is swept across the tops of the plants about 8 inches deep. In alfalfa, a 180° sideways sweep is commonly taken 8 inches deep into the plants, as shown in Figure 7-6.

Another common sampling technique is the use of a *drop sheet* (Fig-

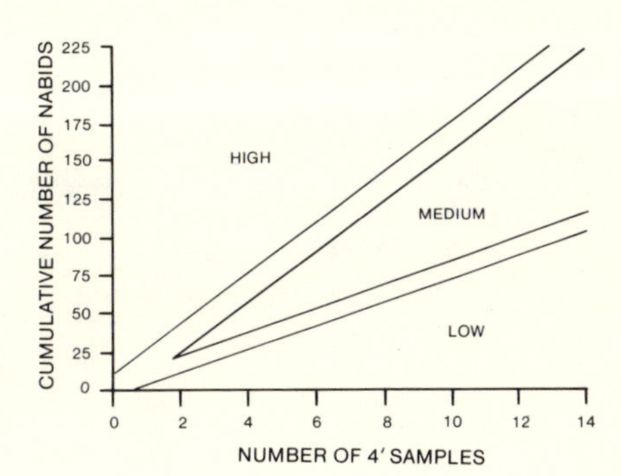

FIGURE 7-5. The sequential sampling model for *Nabis* spp. from which Table 7-1 was derived (from Waddill *et al.*, 1974).

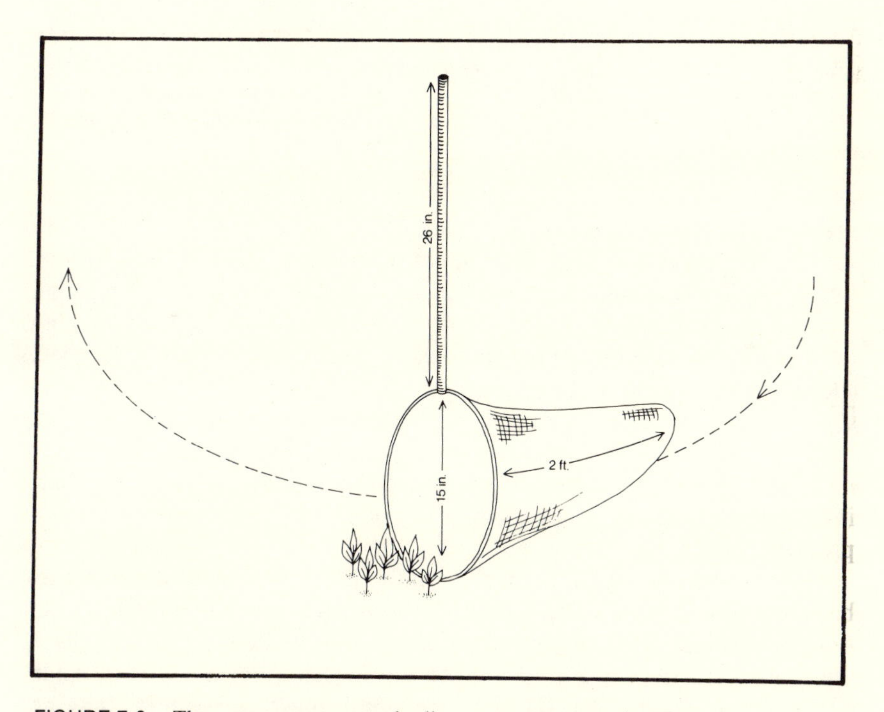

FIGURE 7-6. The proper way to standardize a sweep net sample. The sweep net is swung in a manner describing a 180° arc (half circle) while positioning the open end of the bag vertically 8–10 inches down into the top of the plants.

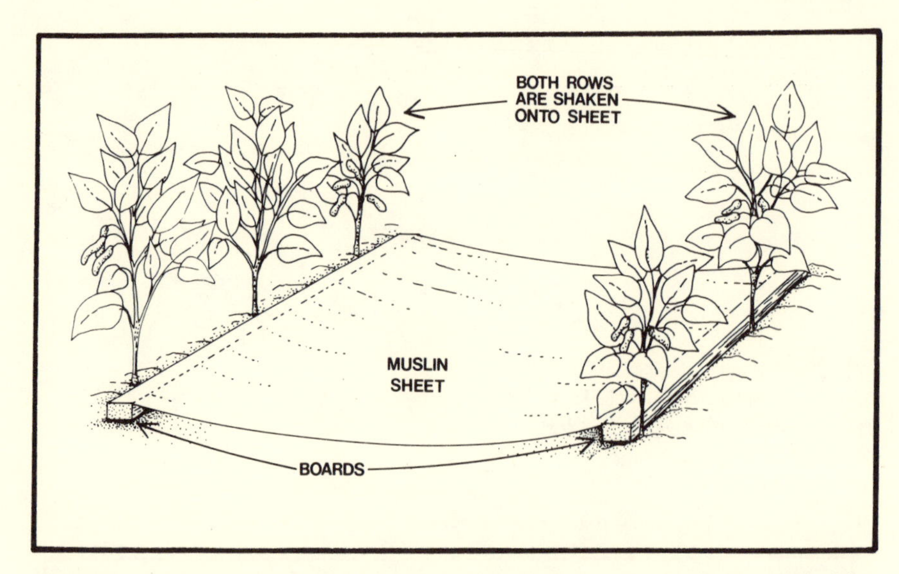

FIGURE 7-7. Sampling with the drop sheet shake method. A muslin sheet, stretched between two boards of a specified length, is placed between rows of plants, and foliage of plants from either side of the row is shaken onto the sheet. Insects drop from plants onto sheet and are counted.

ure 7-7). Nonflying insects are dislodged by shaking 12–18 inches of a row of crop plants onto a muslin sheet placed between the rows beneath the plants. This has been particularly useful in crops where sweeping might damage the crop plants or where insects are situated too low on the plant to be picked up in a sweep net sample. This sampling technique has been successfully used in conjunction with the sequential sampling tables in Figures 7-4 and 7-5 for determining infestations of the green cloverworm in soybeans. It has also been used for sampling leaf feeding caterpillars in cotton. Another sampling tool, the D-Vac, which literally vacuums insects out of the field, is expected to come into wider scale use as improvements are made in its portability.

Many factors can influence sampling results. Several have already been mentioned. One is the *lack of uniformity* between individuals in their sampling technique. Another is the *time of day* when sampling takes place. Many insects are only active and present in the upper areas of the plant during certain periods of the day (e.g., morning or evening). Many are driven lower or out of the plant by rain, wind, or sun. *Size and stage* of the pest commonly influence the numbers observed or taken by a sampling method. Noting the stage of the sampled insect is important in tabulating results and evaluating overall impact, and in predicting when trouble may arise.

The pest manager must recognize all these shortcomings in a potential sampling system and design his own program to avoid as many of these problems as possible. This may require some experimentation and some creative thinking. Integrated pest management is in its infancy; its future and success will be largely determined by the innovation, independence, dedication, and sound scientific approach of today's pest managers.

CONTROL ACTION LEVELS AND OTHER DECISION-MAKING GUIDES

There is, of course, little point in sampling pest populations in a resource management system unless these population densities can somehow be meaningfully related to potential pest damage. Although the mere presence of a pest is enough to cause some growers to apply a pesticide in their crops or to their livestock, integrated pest management programs recognize that in many cases unwarranted pesticide applications can cause more severe problems than the ones they are aimed at solving, or the attendant costs may be greater than the benefits. Accordingly, every effort is made to avoid these unnecessary treatments.

Control action thresholds such as economic or aesthetic injury thresholds have been discussed in earlier chapters (Chapters 3 and 5). The fundamental reason for determining such thresholds is to differentiate between the mere presence or innocuous numbers of a pest and its occurrence in densities high enough to cause significnt damage.

The difference between sampling for mere pest presence and using IPM control action criteria can be best illustrated by drawing on an example from the field: control of the sugarcane borer in the southeastern U.S. In the old system, sugarcane fields were sampled from early spring for larval feeding signs or sugarcane borer-killed tillers. Once an arbitrary number of these superficial signs was observed, the grower was instructed to apply insecticides on a weekly basis for the remainder of the season; no further sampling was carried out. There were several problems with this system. In this case, measuring damage signs rather than live larvae made it impossible for the grower to know whether damaging populations were actually still present, or whether they might have been killed by some natural mortality factor (e.g., natural enemies, severe weather). In addition, when using the automatic spray schedule, the grower had no idea whether damaging numbers of the pests might have continued in the field during the rest of the seaon, or whether his spray applications were just a waste of money. It has also since been discovered that different sugarcane varieties show a marked difference in susceptibility to borer damage; the old pest control system made no allowances for this. This pest control system was further troubled by bad timing of applications, poor placement of insecticides, weak formulations, hindrance by climatic conditions (especially rainfall), and by the buildup of resistance to the insecticides in the sugarcane borer populations. None of these factors could be detected or accommodated in the automatic treatment system. The program also neglected to take into account the fact that the first generation (in the spring) and the fourth and fifth generations (in late summer and fall) of the borer did not inflict economic damage on the sugarcane crop under any circumstances. Treatment continued as long as the crop was in the field.

The sampling system for the integrated pest management program in sugarcane is now initiated in late June just prior to the time when borers could become economically damaging (i.e., when sugarcane stalk internodes form above the soil surface and the second borer generation hatches). Counts are then made weekly of 50 plants 3 feet apart at 6 locations in each 40- to 100-acre field of sugarcane. Treatment is recommended only if infestation attains the 5% level. Retreatment is recommended only after the borer population again reaches a 5% infestation. The 5% economic threshold was derived from the results of large-plot field experiments that correlated yield reductions and infestation levels in conjunction with an estimation of the actual cost of control. The fourth and fifth generations are not treated. This new sampling program combined with the use of a carefully derived economic threshold has saved sugarcane growers much money and many pesticide applications, and slowed down the development of insecticide resistance in their key pest. Moreover, it has reduced the insecticide load used in the American Southeast, an important wildfowl flyway, a factor of great economic as well as aesthetic value.

While the need for developing useful control action or economic thresholds such as the one in sugarcane has been long recognized, surprisingly few exist for the world's major pests. In fact, a recent survey of various University of California insect pest control recommendation circulars revealed the following assortment of nebulous and basically useless action criteria:

when damaging plants	before damage occurs
when present	anytime when present
when damage occurs	early, mid, and late season
when they first appear	on small plants as needed
when colonies easily found	when present and injuring the plants
when abundant	when feeding on the pods
when needed	throughout the season
early season	when infestation spotty
when present in large numbers	when plants are three-feet tall

Many factors have contributed to the lack of useful control action thresholds. The first is the difficulty in establishing a meaningful threshold. The relationship between pest density and yield or pest density and crop revenue is not a linear one (Figure 7-8). In some cases, particularly with insects not attacking the actual crop product, very dense populations can be tolerated with no significant loss in yield. However, this fact has often been very difficult for growers and pest control advisors to accept. And particularly when a crop is attacked by a complex of different pests, it is difficult to know if the effects of combined populations are additive, synergistic, or antagonistic. In fact, research in this area is almost totally lacking.

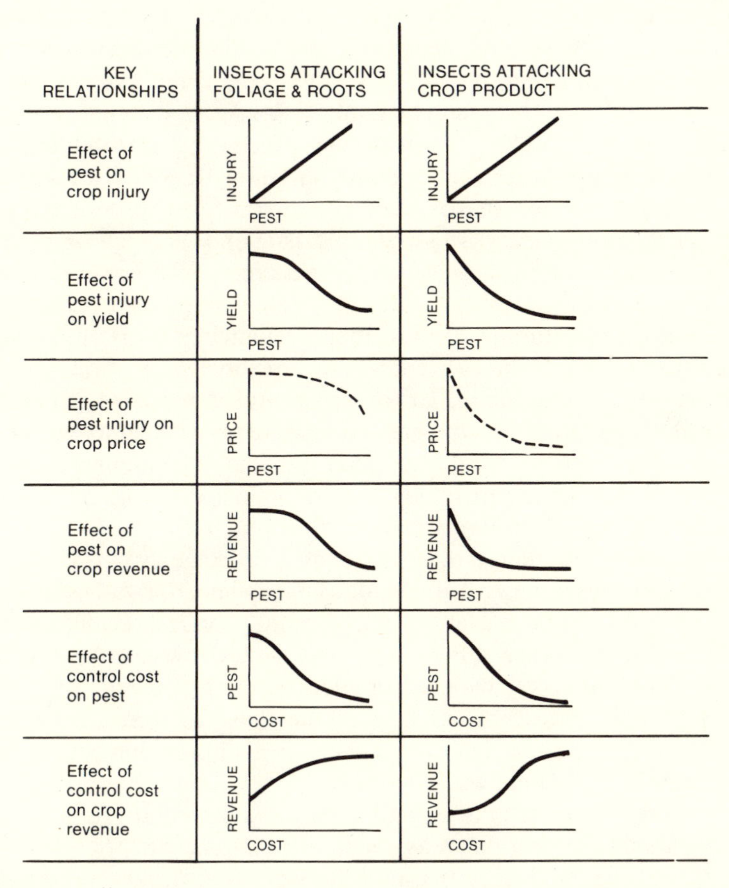

FIGURE 7-8. Generalized relationships between pest density and different types of crop "loss" in a fruit or seed crop. Note the difference in these relationships between insects attacking the marketable crop (direct damage) and those causing indirect damage by attacking roots or leaves (after Southwood and Norton, 1973).

Economic thresholds change throughout the year at different stages of crop development. For instance, heavy tillering rice varieties in Japan can tolerate moderate levels of rice stem borer during early stages of crop growth because lost tillers are rapidly replaced at this point in the crop's life cycle. In fact, in certain crops (e.g., field beans, turnips, wheat) compensatory growth can be so effective that low infestations of some pests early in the season actually result in an increase in yield. On the other end of the scale are pests, such as the tobacco budworm, that can cause severe economic damage in tobacco early in the season but none after all harvestable leafbuds have opened.

Economic thresholds also vary from variety to variety. For instance, irrigated late maturing varieties of grapes, especially table grape varieties in the San Joaquin Valley of California, often require an extra treatment for a "third" brood of leafhoppers, which causes little or no problems in grapes used for wine or raisins or varieties that are harvested earlier. Many semiresistant plant varieties have extremely high economic thresholds as compared to their nonresistant relatives.

Control action thresholds are often affected by unstable marketing standards. For instance, in sweet corn, marketing standards are very strict during the peak of the season, and the tolerance for corn earworms is almost zero. However, when sweet corn is in short supply—e.g., before or after the peak harvest—standards are relaxed, and a few earworms are allowed through.

In untreated cotton in the San Joaquin Valley of California, the bollworm (*Heliothis zea*) has an economic threshold of 15 first or second instar larvae per 100 plants. However, in insecticide-treated fields, the economic threshold drops to eight such larvae per 100 plants because the insecticide's treatment will have killed off the natural enemies of the bollworm. Thus economic thresholds must be adjusted according to the disturbances caused by previous management practices.

In practice, economic thresholds must be constantly revised to account for new pests, new varieties, new management practices, new marketing standards, and variation in commodity prices. Understandably, economic thresholds are at first commonly set too low to offset the logical fear of unknown risks; with experience and experimentation they can usually be adjusted upwards. One of the most comprehensive sets of economic thresholds has been established in California hay alfalfa. These action thresholds are shown in Table 7-2.

Some resources are not managed for direct monetary returns. These include park and recreation areas, mosquito abatement districts, backyard gardens, pantries and kitchens, and various other urban environments. Since an economic injury threshold analysis is not suitable in these sit-

TABLE 7-2
Control Action Thresholds for Insect Pests of Hay Alfalfa as Established by The University of California Division of Agriculture Sciences, January, 1976

Pest	Control action threshold
Spotted alfalfa aphid	Spring: when population level reaches 40 aphids per stem Summer: when population level reaches 20 aphids per stem *Important*: Do not treat first three cuttings if lady beetle/aphid populations are in the following ratio: standing hay—1 adult/ 5–10 aphids or 3 larvae/40 aphids; on stubble—1 larvae/50 aphids Overwintering populations: 50—70 aphids per stem Fall: in reseeded alfalfa in Imperial Valley, 20 aphids per stem
Pea aphid	Treat alfalfa less than 10 inches in height when population level reaches 40 to 50 aphids per stem. In alfalfa 15 inches in height, treat when population reaches 70 to 80 aphids per stem. In alfalfa more than 20 inches in height, treat when population reaches 100 per stem
Alfalfa caterpillar	When an average count of 10 nonparasitized caterpillars per sweep is found in a field of uniform growth
Grasshoppers	At 15 per square yard and dependent on growth stage of hay
Western yellow-striped armyworm	At 15 per sweep when worms are at least one-half inch in length. (Control action recommended at this time is to cut alfalfa and place a barrier of 5% Dylox around the field to prevent migration; field treatment with insecticide not recommended.)
Alfalfa weevil larvae or Egyptian alfalfa weevil larvae	After larvae begin to appear, fields should be checked at 2- to 4-day intervals. Fields should be treated when larval count reaches an average of 20 per sweep On short alfalfa early in the season or on stubble following cutting that cannot be checked with a sweep net, treatment is indicated when growth is retarded because of weevil feeding
Spider mites	Chemical control never recommended; however, if infestation severe, cutting may help avoid loss of leaves
Beet armyworm	At 15 per sweep when worms are at least one-half inch in length

uations, aesthetic injury levels and aesthetic injury thresholds must be developed for these resources. The aesthetic injury level is that level of pest presence or damage which significantly offends the aesthetic values of the people who use the resource. Because of the differing aesthetic values among the human population in any community, such levels may be difficult to establish. Much of the work of the integrated pest management specialist in these areas must be devoted to increasing the community's tolerance of certain organisms whose presence threatens neither the community's health nor the long-term value of the resource (e.g., trees, buildings). Only through proper education of the general public can the pest manager hope that unnecessary pesticide applications will not be made.

Use of Models in IPM Decision-Making

A generalized representation of a pest management system can be diagrammed as in Figure 7-9. The diagram fits equally well all types of pest management systems. The success of each system, however, de-

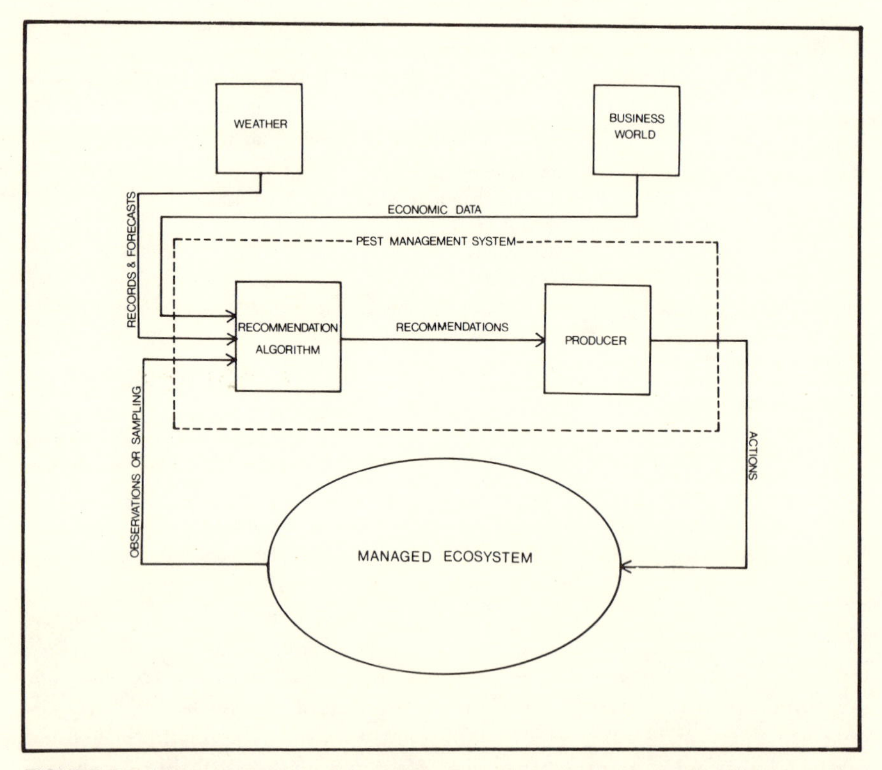

FIGURE 7-9. Diagrammatic representation of a pest management system. The key to the viability of the system is in the recommendation algorithm, i.e., what kinds of information are used, and how they are interpreted in coming to a pest management decision (after Ruesink, 1975).

pends on the appropriateness and completeness of the "recommendation algorithm," i.e., the decision-making process: What factors are considered and how are they considered when making a decision?

In the simplest systems the grower merely uses his intuition and experience together with his own limited and arbitrary field observations in determining need for and choice of control action options. In a more sophisticated system the grower seeks outside advice—e.g., from an extension agent, a pest control consultant, or a published bulletin on his crop. In this kind of system, especially if it involves a well-founded economic threshold, the recommendation algorithm becomes more complex, has a much more complex knowledge–information base, and is better able to interpret and utilize weather and economic and sampling inputs than a recommendation algorithm involving only the farmer's experience and intuition alone. Sometimes, however, the recommendation algorithm contains faulty assumptions about pest or crop biology or the impact of a control action; in these cases, even with good sampling, weather prediction and economic data may be interpreted wrongly, and a poor control action recommendation will result.

Future pest management systems are apt to rely more strongly on mathematical and computer models in their recommendation algorithms. Computer models simulating actual crop and/or pest situations are already being developed for some of our major crops. These include cotton, alfalfa, apple, soybean, and cereals. Such models can embrace far more interactions and options than the grower (or the extension agent or other pest control advisor) is able to handle alone and are the only practical way of dealing with the multitude of complexities that characterize the typical ecosystem. With a good simulation model a researcher can accurately assess the effect of certain actions and pinpoint critical moments in pest and crop life cycles. Experiments can also be carried out on the simulation model to study the effects of new management actions without the constraints of time, money, and disruption that often preclude field experimentation on such a scale.

Two kinds of computer models can be useful to the pest manager. One type uses the computer as a communication system, rapidly rewriting extension bulletins and putting out current information on weather and area-wide pests. Readouts from these models (perhaps piped into a county agent's office) help the grower consider a broad spectrum of options and use a maximum amount and variety of information in making pest control decisions (Figure 7-10). The other type of model takes in information from the individual grower or field (e.g., sampling results, management practices), puts it together with current information from outside the ecosystem (e.g., current prices, weather) and an already established ecosystem

```
*******************************************************************************
   06/26/75    MESSAGE FROM    SCHWALLIER       PHIL
TENIFORM LEAFMINOR
     50 PCT IN THE PUPA STAGE AND 50 PCT HAS EMERGED AS ADULTS THIS
     WEEK IN THE GRAND RAPIDS AREA.

SCAB
     MORE IS SHOWING UP IN THIS AREA THIS WEEK.  NOW APPLES ARE
     SHOWING SOME SCAB.

WHITE APPLE LEAFHOPPER
     FINDING 1ST INSTAR NYMPHS HATCHING AGAIN FROM OVER-
     WINTERING EGGS.  FENNVILLE SAYS ALL THE OVERWINTERING EGGS
     HAVEN'T HATCHED YET, AND ARE STILL VIABLE AND ARE DEVELOPING.
     MOST ALL LEAFHOPPERS WE SEE ARE EITHER THESE 1ST INSTAR
     NYMPHS OR LAST INSTAR NYMPHS OR ADULTS, THE MAJORITY
     ARE ADULTS.

CODLING MOTH
     ENTRIES IN ABANDONED ORCHARD -- VERY COMMON.

MITES
     MOST ORCHARDS HAVE THEM BUT STILL IN LOW NUMBERS.
     HIGHEST COUNT THIS WEEK  6.5 DESTRUCTIVE
     .8 PREDATORY.

     RUST MITES IN 2 ORCHARDS ARE VERY NEAR NECESSARY CONTROL
     HIGHEST COUNT 231 PER LEAF.

APPLE MAGGOT AND RED BANDED LEAFROLLER TRAPS ARE BEING SET THIS
WEEK IN GR AREA.  WE ARE TAKING LAST YEARS TRAPS AND NIPPING THEM
OPEN SO A STICKY SIDE IS SHOWING TOWARD THE OUTSIDE OF THE
TREE.  WE ARE FASTENING THE BAIT TO IT FOR NOW.

SURVEY FROM JUNE 19 THRU JUNE 25

MITES                 75 PCT OF ORCHARDS
GREEN APHID           34  '   '      '
WALH                  50  '   '      '     MOSTLY ADULTS
SCAB
  LEAVES              27  '   '      '
  FRUIT                6  '   '      '
FIREBLIGHT             4  '   '      '

TRAP COUNTS
     CODLING MOTH                   7.2/TRAP/WEEK
     TUFTED APPLE BUDMOTH           6.7/TRAP/WEEK
     ORIENTAL FRUIT MOTH              0/TRAP/WEEK

*******************************************************************************
                      ATTENTION
GUS HOWITT REPORTED HE HAS CAUGHT 1 APPLE MAGGOT IN THREE
LOCATIONS ON JUNE 24 AND ALSO 2ND GENERATION OF OFM AND RBLR
ADULTS ON JUNE 24.
```

FIGURE 7-10. A sample pest alert computer output (for an apple pest management program) showing how current information about area-wide pest populations can be rapidly relayed to growers and pest managers on a regular (e.g., daily or weekly) basis for their use in making pest management decisions. Other information on such computer printouts might include natural enemy population counts, weather predictions, market prices, and suggested management options (from Croft *et al.*, 1976).

information–knowledge base, and is able to suggest optimal control and management strategies and actions to the grower.

INTEGRATED PEST MANAGEMENT OPTIONS

Integrated pest management programs employ a variety of tactics, in concert and in the least disruptive manner possible, to obtain effective, long-lasting pest control. These tactics can be roughly categorized in six classes: biological control, host resistance, autocidal control, cultural control, physical and mechanical control, and chemical control. Each has played an important role in currently existing IPM programs.

Biological Control

When setting up an IPM program or choosing a pest control action, one of the pest manager's first undertakings should be an assessment of the current as well as the potential role of natural enemies in the control of pests in the managed ecosystem. Control by natural enemies (*biological control*) is cheap, effective, "permanent," and nondisruptive of other elements of the ecosystem; as such, it should be a bastion in the pest manager's first line of defense. Unfortunately, it is also the factor most likely to be disturbed by the employment of other pest control tactics, especially after use of pesticides. Use of the more disruptive control tactics should be preceded by a close look at whether or not they will complement rather than inhibit or destroy already functioning biological control agents. This is a key aspect of integrated pest management. Table 7-3 shows some of the advantages biological control has over chemical control, a frequently disruptive control tactic.

Biological control can be considered in three aspects: (1) existing, naturally occurring biological control, (2) the classic importation of natural enemies, or (3) enhancement of the environment to increase the effectiveness of natural enemies.

Every organism is affected by natural suppressive factors operating in its ecosystem: this is basic ecology. The impact of these natural controls is the reason why populations of most of the countless millions of organisms on earth rarely occur in remarkable or destructive numbers. We have discussed the components of natural control in Chapter 3 (see Figure 3-2). In managed ecosystems where food is plentiful and other conditions also ideal for many potential pest organisms, natural enemies can be and, in fact, usually are key factors in keeping these populations well below "pest" levels when undisturbed. However, the importance of this naturally occurring biological control is rarely appreciated until

TABLE 7-3
A Comparison of the Advantages and Disadvantages of Two Pest Control Tactics: Biological and Chemical[a]

Category	Biological control	Chemical control
Environmental pollution; danger to man, wildlife, other nontarget organisms, soil, etc.	None	Considerable
Upsets in natural balance and other ecological disruptions	None	Common
Permanency of control	Permanent	Temporary—must repeat one to many times annually
Development of resistance to the mortality factor	Extremely rare, if ever	Common
General applicability to broad-spectrum pest control	Theoretically unlimited but not expected to apply to all pests. Still underdeveloped. Initial control may take 1–2 years, but then pest remains reduced	Applies empirically to nearly all insects but not satisfactory with some. Can rapidly reduce outbreaks, but they rebound. Psychologically satisfying to the user at first.

[a] From DeBach (1974).

natural enemies are destroyed (often by heavy or repeated pesticide applications) and "new" pests suddenly rise to injurious abundance (*secondary pest outbreaks*, see Figure 7-11), or a major pest becomes even more damaging (*target pest resurgence*, see Chapter 4 and Figure 4-12).

It was the steadily increasing number of such pesticide-induced secondary and resurgent pests and repeated breakdown of the conventional chemical control tactic that caused pest control specialists around the world to look for a new pest management strategy (IPM) that could reduce the occurrence of such outbreaks by utilizing the already available natural control factors in each ecosystem and supplement their action by use of chemicals and other artificial tactics in ways that do not impede control by natural enemies. Well-known examples of minor pests being released from naturally occurring biological control to become pests of major importance include the spider mites on many crops and *Heliothis* spp., especially in cotton. In the case of the spider mites, excessive pesticide use caused the transformation of this group of tiny arthropods from the status of a relatively minor group to number one pest status across a spectrum of crops worldwide in just 10 or 15 years (the late 1940s through the 1950s). Bollworms and budworms (*Heliothis* spp.), once minor pests in the U.S.

cotton belt, are now considered the major problem in many cotton growing areas as a result of their population explosion following the destruction of their natural enemies and development of resistance to many control materials.

It is critical that the integrated pest management specialist appreciate the role of natural enemies in the control or repression of both pest and potential pest organisms in the ecosystem, and that he, wherever possible, augment and encourage these beneficial organisms and employ pest management tactics that enhance rather than impair, their action.

Classic biological control involves the deliberate introduction and establishment of natural enemies into areas where they did not previously occur. Classic biological control programs are employed largely against pests of exotic origin. These are pests that, having become accidentally established in new areas without their key natural enemies, are able to increase in numbers virtually unchecked until their host plants are dev-

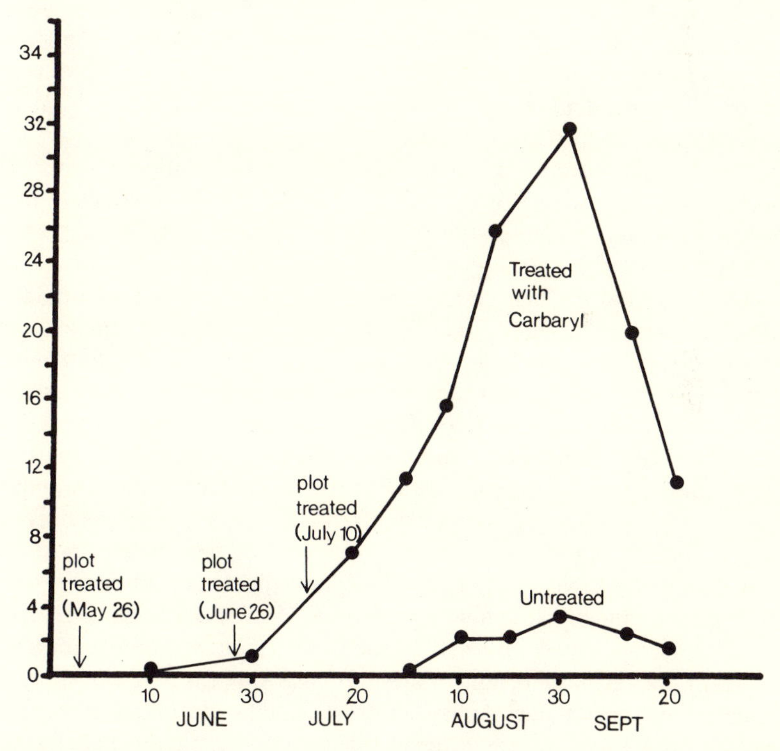

FIGURE 7-11. Secondary pest resurgence. In untreated vineyards the Pacific mite did not reach damaging levels throughout the season; however, in plots sprayed with an insecticide toxic to natural enemies (carbaryl), the Pacific mite rose to damaging levels (from DeBach, 1974).

astated. These same pest species are often relatively unknown (i.e., nonpests) in their native habitats because there they are under effective, naturally occurring biological control. This absence of effective, adapted natural enemies is the reason that species of exotic origin comprise a large percentage of our major pests. Of the 28 most serious pest arthropods in the U.S., 17 (60%) are of exotic origin. Of the 693 weed species treated in *Weeds of California,* 63% are aliens, and these are generally the most serious. Places of origin of most plant pathogen and nematode pests have not been determined, but many of these pests are widespread geographically and may have important natural enemies if their native habitats could be determined.

Classic biological control involves three basic steps. First, the pest's native home must be identified. This, of course, involves a correct prior identification of the pest organism itself. Then a search must be conducted in the native area for natural enemies, and shipments of appropriate species sent back to the newly infected or target habitat. Shipments should include as many individuals as possible, with special attention paid to collecting as climatically well-suited and adaptable a population as possible. After appropriate quarantine processing and biological testing, the primary natural enemies are colonized in the new area. The goal is to reestablish the host–natural enemy relationship and, in doing so, to add another limiting factor to impact upon the pest population and lower its equilibrium level (see Chapter 3, Figure 3-3). Classic biological control has been employed successfully against well over 100 pest insect and weed species worldwide.

The potential for classic biological control is still largely untapped. Financial support has been meager for this control technique, at least partly because in permanently suppressing a pest the effective natural enemy offers no product to market and no long-term profit to the pest control industry. Consequently, classic biological control programs are most invariably conducted by public agencies (e.g., the U.S. Department of Agriculture, certain state departments of agriculture, certain land grant universities) with limited funds that frequently dry up once success is attained. Unfortunately, even growers rapidly forget about a pest once it is successfully biologically controlled, and their support diminishes as they begin to concentrate on new problems. However, the savings in pest control costs and crop losses that such a permanent control technique affords for the resource producer are immense, and these savings increase as each crop season passes. By contrast, chemical control never eliminates a pest (except in rare cases of tentatively established species) and, of course, must be reutilized each year. This represents an everlasting production cost. It is to be hoped that the future will see more public funding and grower support for classic biological control research.

The pest manager may enhance the action of biological control agents (whether naturally occurring or introduced) in the managed ecosystem by employing practices that preserve and augment these beneficial organisms or improve their effectiveness. Such practices include using cultural or other environmental manipulations that will favor natural enemy populations (some of these will be discussed in the next section under cultural controls). The augmentation of food sources for natural enemies has been discussed previously—e.g., the provision of flowering plants as nectar sources for parasites along highway rights-of-way and in apple orchards and planting blackberry bushes to provide an alternative host reservoir for the grape leafhopper parasite. Natural enemy activity can also be enhanced by the provision of artificial food sources such as Wheast®: yeast hydrolysate + sugar + water. As Figure 7-12 illustrates, this practice has been quite successful in increasing predator (green lacewing) populations and fecundity in alfalfa field experiments.

The natural enemies of pests are more numerous in kind than the pests they control. Most cases of classic biological control have involved the importation of insects that are parasites or predators of pest insects or of insect species that feed on weeds. However, birds are important predators in many ecosystems, and they have been effectively managed in European forests with the provision of nesting boxes. Mammals can be important predators as well; the role of cats in limiting numbers of mice and other rodents is one of the many cases of unheralded, naturally occurring biological control. Shrews have been important in Canadian forests in the control of the larch sawfly and other insect pests. Lizards and other small reptiles are important household predators in many tropical countries. Microbial agents (bacteria, viruses and fungi) are extremely important limiting factors in many pest populations. Such microbial agents have been used in classic biological control (e.g., myxomatosis virus was introduced into Australia to control rabbits). A polyhedrosis virus, unintentionally introduced into North America, was able to suppress the European spruce sawfly, while more recently, a rust species of fungus was introduced into Australia, where it controlled the exotic skeleton weed. Naturally occurring epizootics of insect pathogens are recognized as important controls and are regularly monitored in several IPM programs (e.g., in alfalfa in California: polyhedrosis virus of the alfalfa caterpillar, and entomophthoraceous fungi of the spotted alfalfa aphid). Microbial agents are also commercially prepared and applied with application equipment similar to that used for chemical pesticides.

About a dozen microbial agents are under commercial development or in production worldwide. These materials offer an excellent alternative

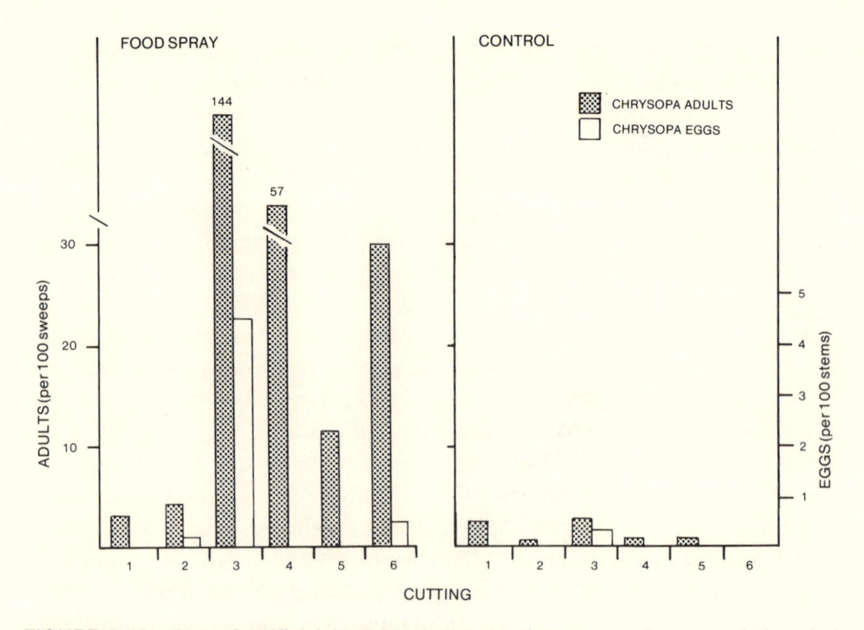

FIGURE 7-12. Use of artificial food supplements to increase predator populations in infested fields. This graph shows the mean number of lacewing (*Chrysopa carnea*) adults and eggs in an alfalfa plot sprayed with food supplements and an unsprayed field. Samples were taken in 1967. The food supplement consisted of yeast hydrolysate + sugar + water and was applied once a week (from Hagen *et al.*, 1970).

to traditional pesticides for IPM programs. Since they usually only infect specific groups of insects, they often cause no disruption of naturally occurring biological control. *Bacillus thuringiensis,* a bacterial pathogen infecting a broad spectrum of insect pests, is the most common microbial insecticide in use today. Milky spore disease called by *Bacillus popilliae* was important in the control of the Japanese bettle in the eastern U.S. in the 1940s and 1950s and is still used today. A commercial preparation of a polyhedrosis virus that kills the cotton bollworm is currently being used on a limited scale in cotton and other crops.

Mass propagation and inundative releases of other biological control agents have been employed with only limited success in the U.S. Somewhat more successful elsewhere, these programs have usually involved general predators such as lacewings or rather polyphagous parasites such as *Trichogramma*. However, outstanding success has been attained with recurrent colonization of certain lady beetles and parasitic wasps against some scale and mealy bug pests of citrus in Southern California and in the use of predatory mites and parasitic wasps against spider mites and aphids, respectively, in glasshouses in the U.K. and mainland Europe. As rearing and release techniques improve, use of inundative biological control may increase. While these methods have the same advantage of being safe from an environmental and health viewpoint, their use will always be considerably more expensive than classical biological control.

Host Resistance

Control of pests by genetic manipulation takes two forms: (1) the manipulation of the genetic makeup of the host plant or animal so that it is resistant to pest attack (*host resistance*) and (2) the manipulation of the genetic makeup of the pest so that it cannot survive in the resource environment (*"autocidal" techniques*). Autocidal control will be discussed in the next section.

The development of host resistance to pests, particularly in agricultural crop plants, has been one of the most successful and ecologically sound techniques used against plant pathogen and nematode pests. Host resistance has also been successfully employed against a number of insect pests (e.g., the Hessian fly, spotted alfalfa aphid, grape phyloxera, woolly apple aphid, corn earworm, and rice stem borer).

Host plant resistance may be due to physiological factors (*antibiosis*; e.g., toxic compounds within plant tissues that inhibit the pest) or mechanical factors (*morphological resistance*; e.g., a cuticle that is too tough for the pest to penetrate or pubescent leaves that inhibit pest adhesion or mobility), or the host plant may actually continue to support pest populations but remain tolerant of pest damage (*tolerance*).

Host resistance is usually most effective and long-lasting if it relies on more than one gene and, preferably, more than one character for its mechanism of resistance. Such "polygenic" resistance makes it more difficult for the pest to develop strains (or *biotypes*) able to overcome the plant's resistance.

Host resistance can come from many sources. In some cases, as with American grapevine resistance to grape phyloxera (see Chapter 4), another strain or variety of host plant has a natural resistance to specific pests. These strains can be interbred with local susceptible varieties (or, as with the phyloxera-resistant grapevines, used as root stock onto which susceptible varieties can be grafted) thus imparting resistance to susceptible varieties in the area of pest infestation. Often closely related wild plant species possess resistance to pests, and these may be hybridized with the cultivated varieties to obtain resistant strains. In some cases natural mutations or ones artificially induced with ionizing radiation, chemicals, or physical treatments have been useful in the production of pest-resistant host plants.

Host resistance is an ideal tactic for use in integrated pest management programs. After the initial development of resistant varieties, the cost of this tactic is minimal to the grower. Additionally, use of host resistance is environmentally sound, generally causing no major disruptions in the managed ecosystem. In a number of instances, integrated control has been greatly enhanced by the use of a low to moderate level of host resistance in conjunction with the control given by natural enemies. Such a combination of plant resistance and biological control tactics (as suggested in Figure 7-13), by limiting pest populations in two unrelated ways, could reduce selection pressure (as illustrated in Figure 4-11) and consequently decrease the chance of buildup of pest populations able to overcome the host plant's mechanism of resistance.

Several other considerations associated with the utilization of resistant plant varieties should be noted, however. Sometimes resistance to one pest is obtained at the cost of increasing susceptibility to other pests; this is especially true when varieties developed for plant pathogen resistance are not adequately tested for susceptibility to insect pests. Development of resistance in a breeding program is a long-term project, often requiring 3–15 years of concerted effort. This entails a major in-

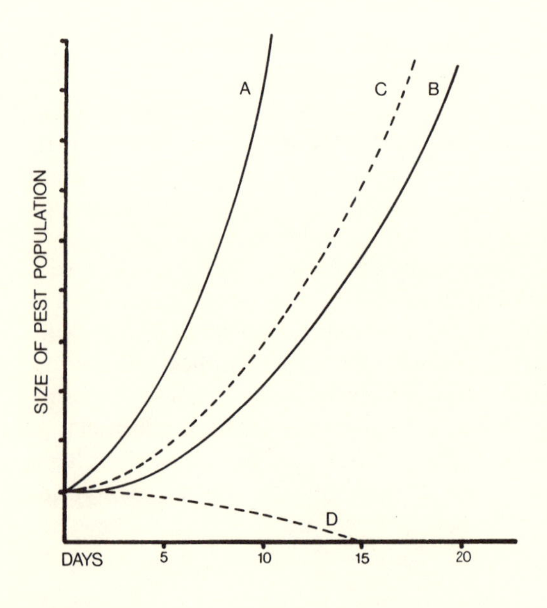

FIGURE 7-13. A hypothetical graph showing the influence of a low level of plant resistance to insect attack on the effectiveness of natural enemies. In this case, a low level of resistance is able to reduce pest populations to a point where natural enemies are able to effectively control pests. This is an excellent example of the integration of host plant resistance and biological control. Solid line, susceptible plant; dashed line, resistant plant. (A) and (C), without predators; (B) and (D), with predators (from van Emden, 1966).

vestment of time and resources; and an alternative control method must be developed to control the pest for the interim period. This last consideration has probably hindered research efforts in this area, especially in the development of resistance against insect pests when chemical control provided quick, immediate results. A third factor which has become a major problem in several programs relying solely on host plant resistance has been the appearance of pest strains able to overcome the host plant's mechanism of resistance. Recent plant breeding programs have attempted to avoid this occurrence by the development of varieties that depend on more than one factor for pest resistance. Despite these problems, it is estimated that about 75% of U.S. cropland utilizes pest resistance varieties, most of these being resistant to plant pathogens. An integrated pest management program that utilizes suppression factors (e.g., natural enemies, cultural controls) in addition to a moderate (rather than total) level of plant resistant will further decrease the possibilities of resistance breakdown.

Autocidal Control

Autocidal controls, as the name implies, involve tactics that cause the pest to contribute to the reduction of its own population. The best known example of this type of tactic has been in the control of the screwworm, a major pest of livestock in the southern and southwestern U.S., through the regular release of laboratory propagated sterile males into the environment.

The principle of the "sterile male" technique of autocidal control is both simple and elegant. If the female of the pest species will mate only once and the laboratory-produced sterile males are equally competitive with wild (or naturally occurring) males as *mates,* mass release of these sterile males (outnumbering native males manyfold) into the environment over time should render large numbers of the native females reproductively impotent and result in a consequent drastic drop in the pest population (Table 7-4). The ultimate goal of this technique is the total eradication of the pest species from an area, and this has occurred with the screwworm on the island of Curacao, in Florida, and in Puerto Rico, and with two species of fruit flies, the melon fly and the Oriental fruit fly, on the island of Rota in the Marianas. However, in Texas, where the screwworm problem is extensive and complicated by the immigration of flies from Mexico, eradication has not been attained. Here, recurrent sterile male releases are made to maintain the wild population at a low level. Recently, this program temporarily broke down, apparently because the "factory"-reared flies developed genetic weaknesses.

TABLE 7-4
Rate of Reduction in a Pest Population with the Sterile Male Technique[a]

Generation	Females (number)	Sterile males (number)	Ratio of sterile to fertile males	Reproducing females (number)
Parental	1,000,000	9,000,000	9:1	100,000
F_1	500,000	9,000,000	18:1	26,316
F_2	131,580	9,000,000	68:1	1,907
F_3	9,535	9,000,000	942:1	10
F_4	50	9,000,000	180,000:1	0

[a] In this hypothetical case, sterile males are released at a 9:1 ratio of sterile males to fertile (natural population) males in the first generation, and releases are continued until eradication is effected.

Sterile male releases have also been employed as preventive measures against "invading" pest species. For instance, continuous releases of sterile males of the Mexican fruit fly are carried out along the California–Mexico border to prevent the entry of the pest into California. A similar program has been employed to prevent pink bollworm moths originating from the desert valleys of Southern California and Arizona from establishing a foothold in the major cotton growing area of California's San Joaquin Valley. The value of these preventative programs has not been experimentally demonstrated, and some entomologists believe that the pests are biologically unsuited for survival in the "protected" areas, explaining their inability to "take root." Further research is needed in these areas.

The autocidal technique is ingenious, has been widely publicized, and has caught many people's imaginations. However, its potential uses are limited, and its role in integrated pest management programs will also probably be limited. Several factors are essential for successful pest eradication by the sterile male technique: (1) the females must mate only a few times, preferably once; (2) the area of release must be geographically isolated so that migrants of the species will not be able to come in from untreated areas; (3) the sterilized males must be sexually competitive with the naturally occurring males; (4) the pest must be amenable to laboratory rearing and such rearing must not have significant effect on its field performance or survival; and (5) there must be an extremely large budget to carry out the program.

The screwworm is probably one of the few major pests of the continental U.S. that satisfies all these requirements. Yet as mentioned above, even the program against this pest suffered a serious setback, owing largely to factors (3) and/or (4) above—an inability of the laboratory-reared males to continue to be sexually competitive and to survive field conditions.

The philosophy of integrated pest management focuses on lowering pest equilibrium levels and combining control techniques to enhance naturally occurring biological control—not on eradicating pests. Consequently, autocidal control has not been an important component in any IPM program. Other autocidal control tactics—e.g., use of hybrid sterility, cytoplasmic incompatibility, and conditional lethals—are still in the basic research phase and are unlikely to be useful in any pest control programs in the near future. However, research on these methods should be encouraged and financially supported.

Cultural Controls

Cultural controls are modifications of management practices that make the environment less favorable to pest reproduction, dispersal, and/or survival. Cultural controls include some of the oldest pest control practices known. Some of these practices, like various kinds of cultivation techniques, have become such habits among many farmers, foresters, and other resource managers that they are often not even recognized as control tactics. Others were abandoned with the arrival of the modern pesticide era because they seemed an unnecessary chore in light of the seemingly "complete" control afforded by the new chemicals alone. However, in many cases, if the growers and other resource managers had kept up the "second line of defense" contained in cultural controls and integrated their use with the new chemicals, selection for pesticide resistance would not have been so common, rigorous, and rapid. For this reason and because chemicals in many situations no longer give complete, effective, and inexpensive control, and have been shown to have adverse effects on natural enemies, wildlife, livestock, and humans, cultural controls are again playing an important role in the management of many pests.

The design and implementation of cultural control tactics require a good knowledge of crop and pest biology, ecology, and phenology, with special attention given to recognizing the "weak links" in the pests' life cycles. Such tactics are generally designed to prevent pest buildup rather than relieve an already existing pest problem; accordingly, timing is critical to the success of many cultural controls. As they are usually simply modifications of operations that would be carried out in any case, the cost of their incorporation into pest management systems is minimal in most cases. Hundreds of cultural control techniques are currently practiced in all areas of pest management; a few of the most common and potentially useful ones are described below.

Sanitation

Sanitation involves the removal or destruction of breeding, refuge, and overwintering sites of pests. Applications of this method are available for nearly every type of habitat and for most kinds of pests—e.g., various insects, plant pathogens, rodents, and nematodes.

In horticultural and tree fruit crop situations it has been particularly useful against twig and branch-inhabiting pests through the destruction of tree prunings and the removal of dropped fruit that might provide overwintering sites. In forestry, disposal of slash resulting from logging activity may reduce bark bettle infestation.

In annual field crop situations, stalk and other crop debris destruction programs have been especially effective against stalk boring insects and rootworms. For example, in many cotton-growing areas a rigorous and very successful cultural control program has been instigated against the boll weevil and pink bollworm. The program prevents the production of sufficient food for the last (overwintering) generation and in doing so greatly reduces the number of individuals of each of these pests able to survive and infest the cotton fields the following year. Three steps are involved. First, when the cotton is mature, the plants are chemically defoliated. After the cotton is harvested, the cotton stalks are immediately shredded, and the residue is plowed under the top layer of soil. This cultural practice has been well incorporated into IPM programs in cotton (Figure 7-14) because it has no significant effect on the natural enemies of two other cotton pests, the cotton bollworm and the tobacco budworm. The preservation of these natural enemies is important because, if undisturbed, they are able to keep the bollworm and budworm well below economic levels. Some growers have found that a heavy foraging of cotton fields by goats or cattle immediately after harvest may provide better than 94% kill of the pink bollworm. In this case, care must be taken that pesticide residues do not harm the livestock.

Sanitation has been an important component of many plant pathogen control programs. Destruction of tobacco stalks has been an important factor in the control of several diseases of this crop (see Chapter 8). Destruction of volunteer plants and crop-free periods have been useful in reducing plant pathogen damage in several crops, especially in cases where crops are grown all year. These have been important techniques in the control of western celery mosaic in lettuce and other crucifer crops. The incidence of tobacco mosaic virus in the tomato is reduced if workers transplanting seedlings wash their hands, tools, and machinery in a milk solution. Tobacco mosaic virus (TMV) can often be spread from cigarette and other tobacco sources to the young tomato plants if precautions are not followed.

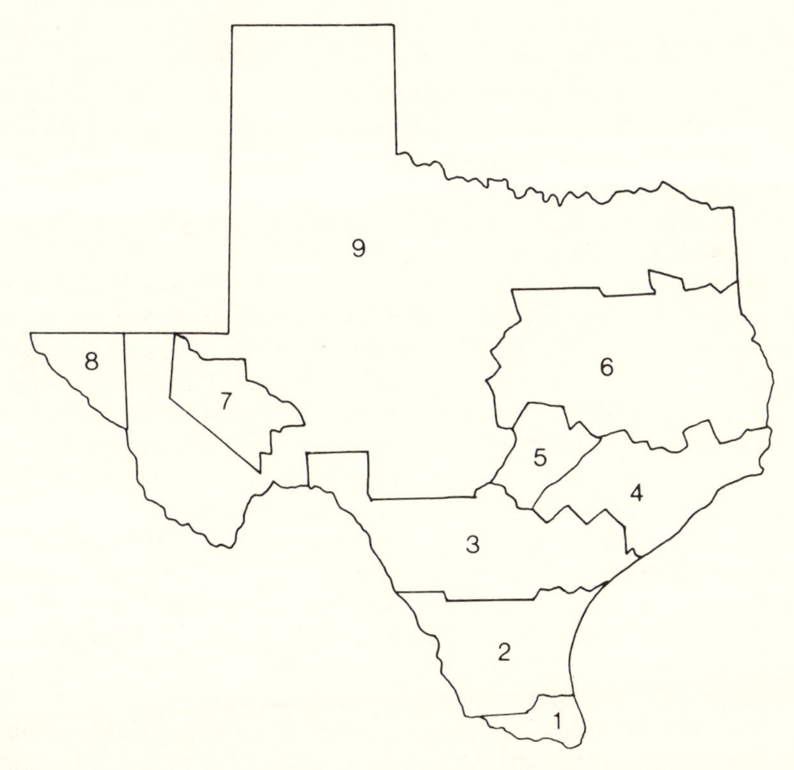

FIGURE 7-14. Cultural control program for pink bollworms in Texas showing legally designated cultural control zones, dates of planting, and stalk destruction deadlines for each zone as set by law. Differences in zone planting dates and stalk destruction deadlines are due to climatic differences in each zone affecting pink bollworm emergence in the spring and overwintering diapause in the fall. Zone 1: Planting period, February 1–March 31; stalk destruction deadline, August 31 (fall okra cannot be planted in this zone before October 15). Zone 2: Planting period, February 15–April 20; stalk destruction deadline, September 25. Zone 3: Planting period, March 5–May 10; stalk destruction deadline, October 10. Zone 4: Planting period, March 10–May 20; stalk destruction deadline, October 20. Zone 5: Planting period, March 10–May 20; stalk destruction deadline, October 31. Zone 6: Planting period, March 20–May 31; stalk destruction deadline, November 30. Zones 7 and 8: Stalk destruction deadline, February 1. Zone 9: No mandatory stalk destruction deadline.

The key to controlling many urban pests such as flies, cockroaches, and rats is to remove sources of food and eliminate their breeding sites—commonly near garbage or other accumulations of moist filth. Mosquitoes often breed in abandoned containers that collect water—e.g., old tires, tin cans, junked cars, old barrels. Removal of these potential breeding sites can greatly reduce urban mosquito problems.

Crop Rotation

Crop rotation is also a well-established "tried and true" control technique. It is especially effective for pests that cannot survive for long periods of time (i.e., one or two seasons) without crop or favored host contact. These include plant pathogens, nematodes, insects, and a number of weeds. This technique is economical and important in the control of many nematode problems.

An enforced rotation program in the Imperial Valley of California has been an effective control for the sugar beet cyst nematode. Under this program, sugar beets may not be grown more than 2 years in succession or more than 4 years out of 10 in clean fields (i.e., noninfested fields), and in infested fields, every year of a sugar beet crop must be followed by three years of a nonhost crop. Other nematode pests commonly controlled with crop rotation methods include the golden nematode of potato, many root knot nematodes, and the soybean cyst nematode.

For many years, a 4-year rotation of corn, oats, and clover in the Midwest kept corn rootworms well below economically damaging levels. However, the arrival of effective soil insecticides made the 4-year rotation unnecessary, and it was abandoned by almost all growers since growing corn every year was more profitable. But the abandonment of the rotation programs led to increased rootworm populations, increased pesticide use, and increased soil pesticide residues. Eventually the rootworm became resistant to many of the soil insecticides. However, for economic reasons, farmers are still unwilling to return to the 4-year rotation.

Cultivation

Until recently, tillage was the only way to control many weed species. This practice is rapidly becoming a thing of the past, however, as more and more growers and gardeners have begun to rely entirely on herbicides and "minimum tillage" techniques, finding them easier and more profitable over the short-run than traditional cultivation practices. In fact, the age-old technique of fallowing land by frequent disking, plowing, and harrowing is rarely practiced on today's modern farms.

Tillage practices can kill pests by mechanical injury, starvation (via debris destruction), desiccation, and exposure. For instance, the wheat stem sawfly in North Dakota can be reduced by as much as 75% by cultivation-caused injury, exposure, and starvation. Summer tillage of fallowed wheat fields, along with destruction of nearby volunteer wheat plants, also destroys wheat streak virus reservoirs and the viruses' vector, the wheat curl mite—thus cleaning up for the next wheat crop. Tillage practices must often be carefully timed to protect natural enemy species and to avoid the lowering of preplant spring soil moisture in nonirrigated fields.

Trap Crops

The practice of attracting pests to a small early crop or alternated crop, which can then be destroyed or sprayed with a heavy dose of pesticide without danger to natural enemies in the major portion of the field (and at a much lower cost), has been employed against parasitic plant, nematode, and insect pests with considerable success.

In cotton-growing areas where the boll weevil is a problem, large numbers of this pest can be attracted to an early fruiting crop (5% of the field). This portion of the field is then sprayed for boll weevil control with no adverse effects on the natural enemies inhabiting the other 95% of the field. If an attractant (such as the pheromone grandlure) is added to the early fruiting crop area, the control may be even greater. In many Texas cotton-growing areas, this practice, along with the sanitation practices mentioned above, has eliminated the need for additional pesticide application for boll weevil control in many fields.

In Hawaii, squash and melon fields are often surrounded by a few rows of corn. These corn plants are very attractive to melon flies, a major pest of cucurbit crops. Generally, treatment of the corn plants controls the flies, leaves no pesticide residues on the melon or squash crops, and is harmless to natural enemies in the crop plants.

Habitat Diversification

Several cases illustrating the beneficial effects of habitat diversification have been discussed in earlier chapters (e.g., strip cutting of alfalfa in Chapter 6, provision of blackberry bush refuges for parasites of the grape leafhopper in Chapter 3). Other examples are common. Leaving a ground cover of johnson or Sudan grass (a minor cultural practice modification) in California's San Joaquin Valley vineyards has resulted in a habitat modification that greatly enhanced the activities of predators of phytophagous mites (see Figure 7-15). Similarily, studies in Europe and Canada have shown that unsprayed orchards with a ground cover abundant in nectar-producing flowers has resulted in significantly higher parasitism of tent caterpillar eggs and codling moth larvae than in similar orchards lacking such flowers.

Habitat diversification is generally utilized as a method to enhance the activities and survival of natural enemies by providing alternate food sources or refuges for these beneficial insects. As has been stressed earlier, adding new elements to the ecosystem may have effects on other pests (e.g., as alternate hosts for plant pathogens). These possibilities must be thoroughly explored before such practices can be made a permanent part of the management system.

Time of Planting

Often the planting of a crop can be timed so the crop gains a competitive advantage over the pest. Several examples will illustrate this common practice.

In wheat-growing areas of the Midwest, Hessian fly adults, which lay the eggs for the fall generation, emerge over a 30-day period during the late summer and survive only a few days after emergence. Since this period is predictable annually, appropriate "safe" planting dates can be established in differing areas after the flies have emerged and died (Table 7-5).

Early planting of corn in certain northern areas of the U.S. can be an effective control for the corn earworm and fall armyworm. These two pests migrate north from overwintering sites in the southern states and do not arrive in the northernmost areas until late in the season. Similarly, early-spring-planted pickling cucumbers often escape damage from the

TABLE 7-5
Average Yields of Wheat and Percentage of Hessian Fly Infestation for 8 Years in Fields Planted before the Safe-Seeding Date[a]

	Average yield		Average percent of infestation	
Location of field	From wheat sown *before* the safe-seeding date (bushels)	From wheat sown *after* the safe-seeding date (bushels)	In wheat sown *before* the safe-seeding date (percent)	In wheat sown *after* the safe-seeding date (percent)
Rockford, Illinois	21.8	28.1	24.5	1.7
Bureau, Illinois	27.4	32.9	45.5	5.2
La Harpe, Illinois	30.8	36.5	38.0	1.8
Urbana, Illinois	29.5	37.1	32.6	5.4
Virden, Illinois	23.6	28.4	48.0	6.3
Centralia, Illinois	14.5	21.9	81.0	8.0
Carbondale, Illinois	21.5	23.9	16.0	1.0
Grand Chain, Illinois	15.5	21.4	32.3	1.0
Average	23.1	28.8	39.7	3.8

[a] From Metcalf and Flint (1951).

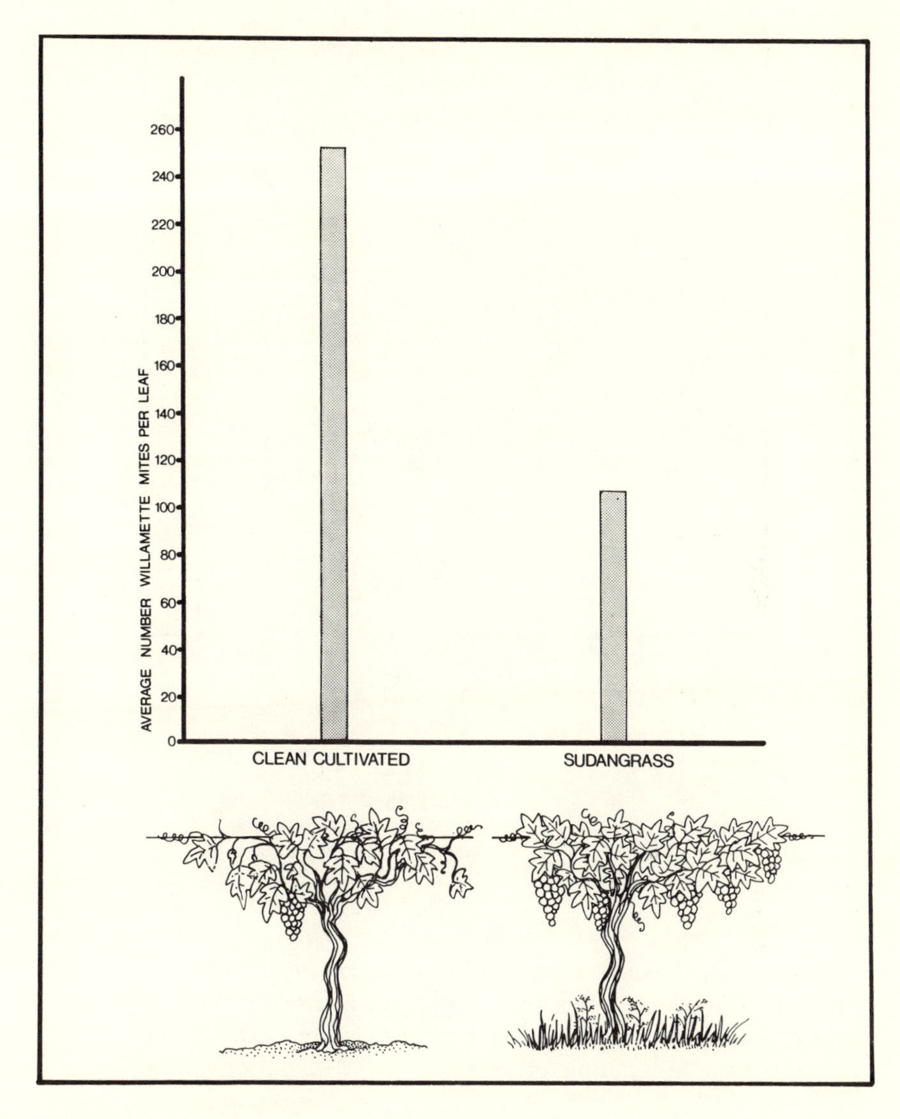

FIGURE 7-15. Effect of ground cover on Willamette mite populations in a vineyard (data from Flaherty *et al.*, 1971).

pickle worm, which heads northward in the summer from overwintering areas in the Gulf Coast States.

Figure 7-16 shows how planting time may be coordinated to avoid adverse environmental factors as well as insect pests and produce a higher crop yield in a sunflower crop. In this case, an early March planting time

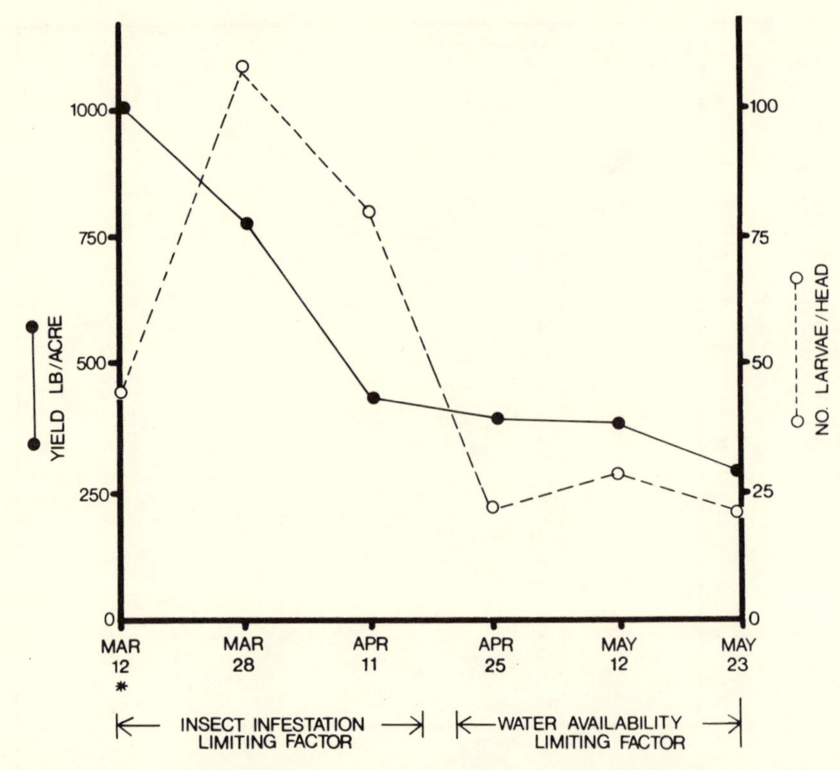

FIGURE 7-16. Effect of planting dates on sunflower moth populations and yield in McGregor, Texas, 1969. Note that later in the season water availability is the limiting factor on yield, not pest numbers. *, Ideal planting date (after Teetes and Randolph, 1971).

gives the greatest yields. Although sunflower moth infestations are lighter in late-April- and May-planted crops, there is a severe loss of yield because of inadequate moisture that more than compensates for the absence of pest damage at this time of year.

Harvesting Practices

In many cases, losses owing to pest damage can be greatly reduced by harvesting as soon as possible, especially when near-damaging infestation levels are noted as the crop reaches maturity. Such "early harvest" practices have been useful against such pests as the sugarcane borer, sweet potato weevil, potato tuberworm, pea weevil, cabbage looper, and imported cabbage worm. Selective harvesting of lumber by a system of early detection of bark beetle damage and removal of infested trees has reduced economic losses considerably in many forest areas.

In alfalfa, a crop that is repeatedly harvested throughout the season, early cutting of the first two harvests can often provide good control of the alfalfa weevil when pest sampling has indicated that economic damage may occur. During the early part of the season the weevil is still in the larval or pupal stages and cannot tolerate the loss of food or physical shock of dryness and heat that occurs when the stalks and leaves of alfalfa are removed in these early cuttings.

Water and Fertilizer Management

At present in U.S. agricultural situations, off-season flooding of fields to kill pests is rarely agronomically or economically sound. On the other hand, a lack of irrigation can cause a threatened crop to mature early and avoid pest damage. Likewise, extra irrigation may extend the growing season of a crop, causing the development of an extra generation of (and extra damage from) a pest; this has been the case with the pink bollworm in cotton where extra irrigation may often add to the next year's problems as well (Figure 7-17). Thus the costs and benefits of extra irrigations must be carefully weighed.

Enhancement of soil fertility affects different crops and different pests in various ways. In some cases, added fertilizer allows the crop to

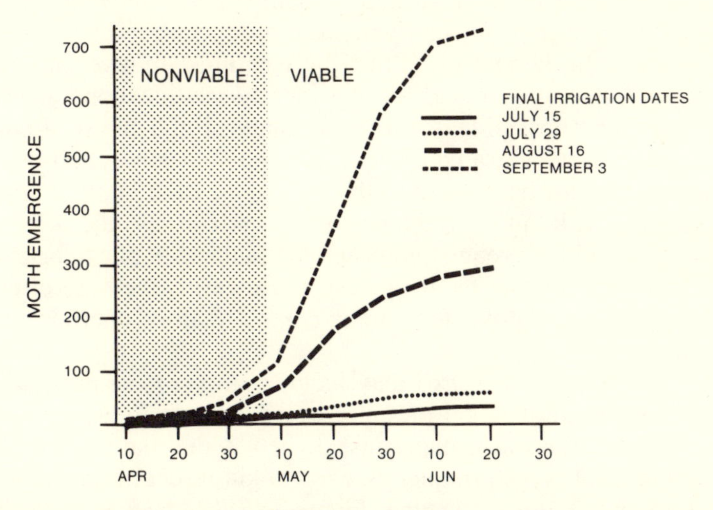

FIGURE 7-17. Effect of final irrigation dates on pink bollworm moth emergence the following year. The later-irrigated fields take longer to mature and thus provide food for the critical overwintering generation of pink bollworms (from Watson *et al.*, 1975).

outcompete weeds (e.g., foxtail in corn); in other cases the addition of fertilizer allows the weeds to gain an advantage over the crop (e.g., wild buckwheat in wheat). Similar instances can be cited for the effect of fertilizer on insect, nematode, and plant pathogen pests. Accordingly, each case must be considered individually and as it affects the overall IPM program.

Use of Pest-Free Seed and Planting Stock

In many crops, nematode- and virus-free transplants or seeding propagules are available; but for one reason or another (often economic) growers do not use them, and these unnecessary pest problems continue to plague certain crops. Examples include sugarcane, sweet potatoes, strawberries, and tobacco. Similarly, pest-free seed is available for many crops.

When contaminated seed and stock are introduced into an area, it often means disaster for neighboring growers as well as the negligent buyer. For this reason some pest managers feel that legal enforcement of these and many other cultural control practices should be required.

Mechanical and Physical Control

Mechanical and physical controls are direct or indirect (nonchemical) measures that destroy pests outright or make the environment unsuitable for their entry, dispersal, survival, or reproduction. They are distinguished from cultural controls in that these actions are taken specifically for pest control purposes and are not merely modifications of existing management practices. Many times, mechanical and physical controls require considerable extra equipment, materials, or labor, and therefore sometimes may not be economically justifiable. Like cultural controls, they exploit "weak links" in the pest's life cycle or specific pest behavioral patterns and thus require considerable biological, phenological, and ecological understanding for their design. However, they are often (but not always) nondisruptive to natural control factors operating in the ecosystem.

Physical control measures may include *temperature manipulation*. Heat and steam sterilization of soil are widely used in greenhouse operations for the control of many insect, nematode, and plant pathogen pests. Steam heat is also frequently used to kill insect pests and mildew in furniture and clothing. Flaming (Figure 7-18) has been used to control weeds in certain situations (e.g., some industrial sites, roadsides, abandoned fields) and is also used as a joint control against weeds and weevils

FIGURE 7-18. Flaming. This physical method of control may be used to control annual weeds along roadsides, or alfalfa weevils in cold regions when plants are dormant. However, because it has a drastic effect (although temporary) on the total ecosystem, all consequences of its use should be considered carefully.

in alfalfa in some areas. Flaming in late fall or winter when alfalfa plants are dormant destroys the weevil adults and eggs by burning the upper portions of plants but leaving the roots undamaged to resprout in the spring. A problem with the flaming method is that it commonly destroys natural enemy and other beneficial species as well as the pests.

Cold storage is used to control many stored-product pests. Many of these pests are of tropical origin and cannot tolerate near-freezing temperatures. Even temperate climate species are often vulnerable to cold temperatures during portions of their life cycle. For instance, the apple maggot and plum curculio in apples can be destroyed by storage at 0°C. Fruit from the northeastern U.S. stored 90 days at this temperature is able to pass the strict California quarantine laws. Drywood termites in furniture can also be effectively controlled by exposure to subfreezing temperatures. Most stored product insect and rodent pests may be controlled by storing food in rooms with little or no oxygen and high CO_2 content.

Although the attraction of insects to lights has been primarily exploited in traps used for sampling purposes, occasionally such *light traps* are used in controlling pests as well. One such case is the use of black-light traps to attract nocturnal insects away from patios, drive-in restaurants, golf driving ranges, and processing plants where insects may contaminate products. The black-light trap is generally placed about 50 feet from the protected area, and yellow (nonattractive) lights are used as illumination for humans. Black-light traps have been incorporated into an integrated control program for the four major insect pests of cigar wrapper tobacco in Florida. In this program insecticide treatments were reduced from 17 to 2 in one season by the use of perimeter-mount black-light traps for the cabbage looper and tobacco budworm. However, because of their expense and high energy requirement, light traps used for control purposes are not expected to become a major component of many integrated pest management programs.

Pest barriers of various types are among the most common physical control methods. The use of ditches and furrows for controlling migrating chinch bugs and armyworms was discussed in Chapter 4. This was a common control method in the U.S. during the early 1900s and continues to be used in many other parts of the world. Metal barriers around buildings are effective deterrents to the subterranean termite.

Screens are probably the best known physical barrier. Screened windows, doors, tents, and other enclosures are effective barriers to flies, mosquitoes, and many other kinds of hazardous or bothersome insects. Screens placed in irrigation pipes and ditches can reduce weed seed dispersal as well as dispersal of snail hosts of the dreaded schistosomiasis organism to newly irrigated regions in Africa and the Near East.

Adhesive substances have provided adequate and environmentally sound control in some instances. Products based on hydrogenated castor oil, natural gum resin, or vegetable wax are common adhesives. Flypaper is a well-known application of this technique. Red "sticky spheres" (Figure 7-3B) have controlled apple maggots in small orchards. Sticky bands (Chapter 4, Figure 4-8) around tree trunks are still a popular control for gypsy moth, cankerworms, cicadas, ants, and several other insect pests of fruit and shade trees.

The simplest mechanical control device is the flyswatter—and nothing quite its equal has ever been invented for the occasional fly in a domestic situation. Other mechanical controls are more complex. Flour mills and many other food processing plants have modern sifting and separating equipment to remove insects and other alien matter. Mechanical delinting of cotton seed greatly reduces pink bollworm problems, and sulfuric acid delinting provides 100% control of seed-borne pink bollworms. Washing trees with water or a mild soap solution has been an effective and safe way of knocking insects off shade trees in urban areas.

Water management (which involves the integration of biological, cultural, physical, and mechanical controls) for mosquito control is a particularly important control method. This includes such operations as periodic flushing, adequate drainage for irrigated pastures, effective management of rice paddies, and avoidance of stagnant water buildup in any other type of water collecting situation from tree crotches to bomb craters. Mosquito management will be discussed more extensively in Chapter 8.

Chemical Control

Chemical pesticides—literally, chemical pest killers—have been developed for just about every group of plants, microorganisms, and animals containing any members that have ever been considered undesirable. They are the most frequently used and among the most useful pest control tools. The present discussion centers on chemical insecticides (and we generally mean also acaricides for mite control) with a brief review of bactericides, fungicides, and herbicides. Chemical materials are also manufactured for the control of rodents, birds, coyotes, nematodes, and many other animals; some of these are the same as or similar to the compounds used in the control of insects; others are quite different. The reader is advised to consult other sources for further details on specific materials.

Any person working with pesticides in the field should be aware of the identity of the chemicals being used; the potential human, livestock, and environmental toxicity of each material; safety precautions that may be needed; and what to do in case of an emergency. It is also important from a pest management point of view to know the identity of the target pest(s) for each application; what nontarget species in the field, especially natural enemies and pollinators, might be affected by the application; and what alternative methods or materials may be available for control of the target pest(s).

The individual chemicals used in pest control may be referred to by one or more of four different classes of names. First is the *common name* chosen by the appropriate scientific society (for instance, the Entomological Society of America, the Weed Science Society of America, or the American Phytopathological Society) and approved by the American National Standards Institute. Second is the *brand (or trade) name*. Since many materials are manufactured by more than one company, one chemical may have from one to six, or even more, trade names. This can be very confusing. To avoid confusion, all pesticides in the following discussion will be described by their common names (although well-known trade names may follow in parentheses). Third is the *chemical name* as a chemist would describe the material. Chemical names are usually too long and cumbersome for everyday use. And last is the *structural formula* that shows the location of the atoms of the pesticide molecule in relation to each other. Figure 7-19 illustrates the four kinds of names for carbaryl, a widely used insecticide.

Over a thousand different materials are registered in the United States as pesticide active ingredients. These in turn are formulated into many times that many commercially available products. Insecticides were the most widely used type of pesticide until 1975. As Figure 1-2 in Chapter 1 shows, in that year the U.S. use of herbicides surpassed the use of

1.	Common Name:	Carbaryl
2.	Trade Name:	Sevin
3.	Chemical Name:	1-naphthyl methylcarbamate
4.	Structural Formula:	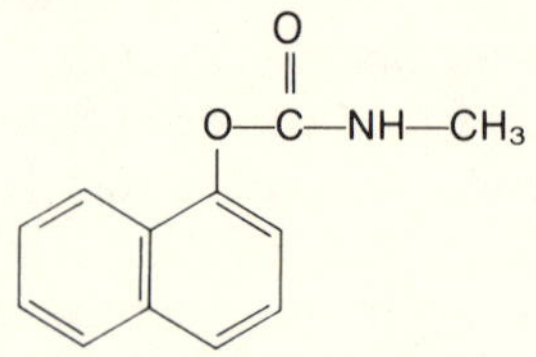 O ‖ O—C—NH—CH_3 (attached to naphthalene ring)

FIGURE 7-19. Nomenclature of pesticides.

insecticides in pounds used annually. Although the production of pesticides continued to rise through the 1950s, 1960s, and 1970s (Figure 7-20), the increase in crop production gained from this increased use of pesticides along with other management innovations was hardly comparable in degree. The increase in pesticide use was in large measure due to merchandising activities of the agrichemical industry, changes in users' perceptions of the need for chemical controls, increases in labor costs (especially for weed control), and increased mechanization of farms and the factors involved in the "pesticide treadmill"—pest resurgence, secondary pest outbreak, and pesticide resistance.

Insecticides and Acaricides

A great variety of pesticide materials have been developed to kill insect and mite pests. Such variety is needed because there is great diversity among insects and mite species in susceptibility to various toxins, and also because in many cases there is a need to selectively kill only certain species, leaving others such as honeybees and natural enemies unharmed. The growing problem of insecticide resistance among insect species has also increased the need for new kinds of materials. Insecticides can be broadly classed according to their chemical nature or their effect on the pest insect. The major groupings are discussed below.

Chlorinated Hydrocarbons. The chlorinated hydrocarbons were the first group of synthetic organic insecticides to appear in the 1950s. Their mode of action is still not completely understood. However, they impair normal transmission of nervous signals, causing nerve impulses to fire spontaneously. This spontaneous firing at first causes the muscles to twitch; the twitching later develops into tremors, convulsions, and eventual nervous and muscular collapse. Most chlorinated hydrocarbons are quite persistent in the environment, that is, they are stable and often do not degrade into less toxic substances for many years (see Table 5-4, Chapter 5).

Chlorinated hydrocarbons in use as insecticides or acaricides include toxaphene, dicofol (Kelthane®), endosulfan, methoxychlor, lindane, pentachlorophenol, and chlordane. A number of these materials, formerly widely used, are no longer registered in the U.S. for most uses because of associated environmental or health problems. These include DDT, aldrin, dieldrin, heptachlor, endrin, Mirex®, and Kepone®, and, for most uses, chlordane.

Organophosphates. The organophosphates were discovered by the Germans in their search for effective human nerve gases in World War

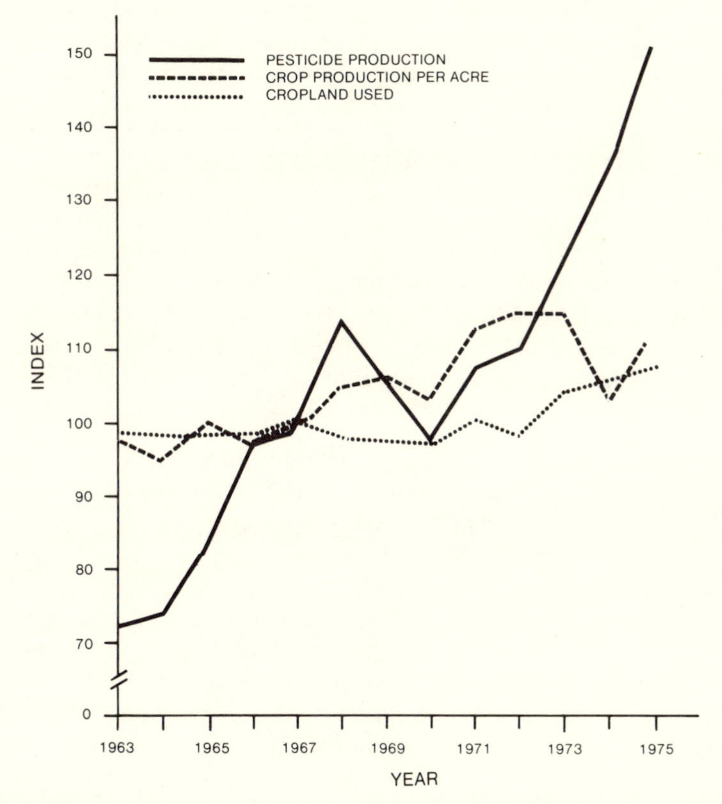

FIGURE 7-20. U.S. production of insecticides and its relationship to acres under cultivation and production per acre.

II. And they are, in fact, nerve gases for both mammals and insects. In general, compared to the chlorinated hydrocarbons, the organophosphates are much more toxic in low doses to insects, people, and other animals and kill a much wider variety of animals, but are less persistent in the environment. Some, however, such as parathion, break down into more toxic materials.

Most of the organophosphates are cholinesterase inhibitors. Cholinesterase is an important enzyme that "turns off" individual impulses after they have been transmitted across a nerve-to-muscle connection. If the action of cholinesterase is inhibited, the impulse remains *on* and the circuit is effectively jammed—keeping the muscle twitching and making it impossible for further messages to get through. Figure 7-21 shows the action of these poisons in more detail. A few organophosphates are also direct nerve poisons; leptophos, for instance, erodes the nerve sheath (lining), causing permanent injury and at times death. Widely used organophosphate insecticides include malathion, trichlorofon, monocrotophos (Azodrin®), mevinphos (Phosdrin®), dimethoate (Cygon®), disulfoton (Di-Syston®), demeton (Systox®), parathion, methyl parathion, diazinon, azinphosmethyl (Guthion®), and dichlorvos (DDVP).

Carbamates. Like the organophosphates, carbamates kill by inhibiting cholinesterase activity. Carbaryl (Sevin®) is the best known carbamate. It is widely used because it has a low toxicity to mammals; however, it kills a wide range of insects—pest and beneficial alike—and can be particularly harmful to bees and parasites. Propoxur (Baygon®) is widely used against cockroaches and other household pests. The plant systemics methomyl (Lannate®), aldicarb (Temik®), and carbofuran (Furadan®) are also carbamates. Not all carbamates share carbaryl's low oral toxicity to mammals; aldicarb and carbofuran, for example, are highly toxic to mammals and birds.

Formamidines. Although this very new and largely undeveloped group of insecticides also acts on the nervous system, the symptoms produced by formamidine-poisoned insects are quite different from those following organophosphate or carbamate poisoning. For this reason it is suspected that their mode of action is also quite different. Formamidines could be especially useful against organophosphate- and carbamate-resistant pests. Examples of formamidine insecticides include formetanate hydrochloride (Carzol®) and chlordimeform (Galecron® or Fundal®). However, the use of chlordimeform has been recently limited by some state regulatory agencies because of suspected carcinogenicity.

Inorganics. Once the most commonly used insecticides, this group has gradually been replaced by the synthetic organic compounds. Many

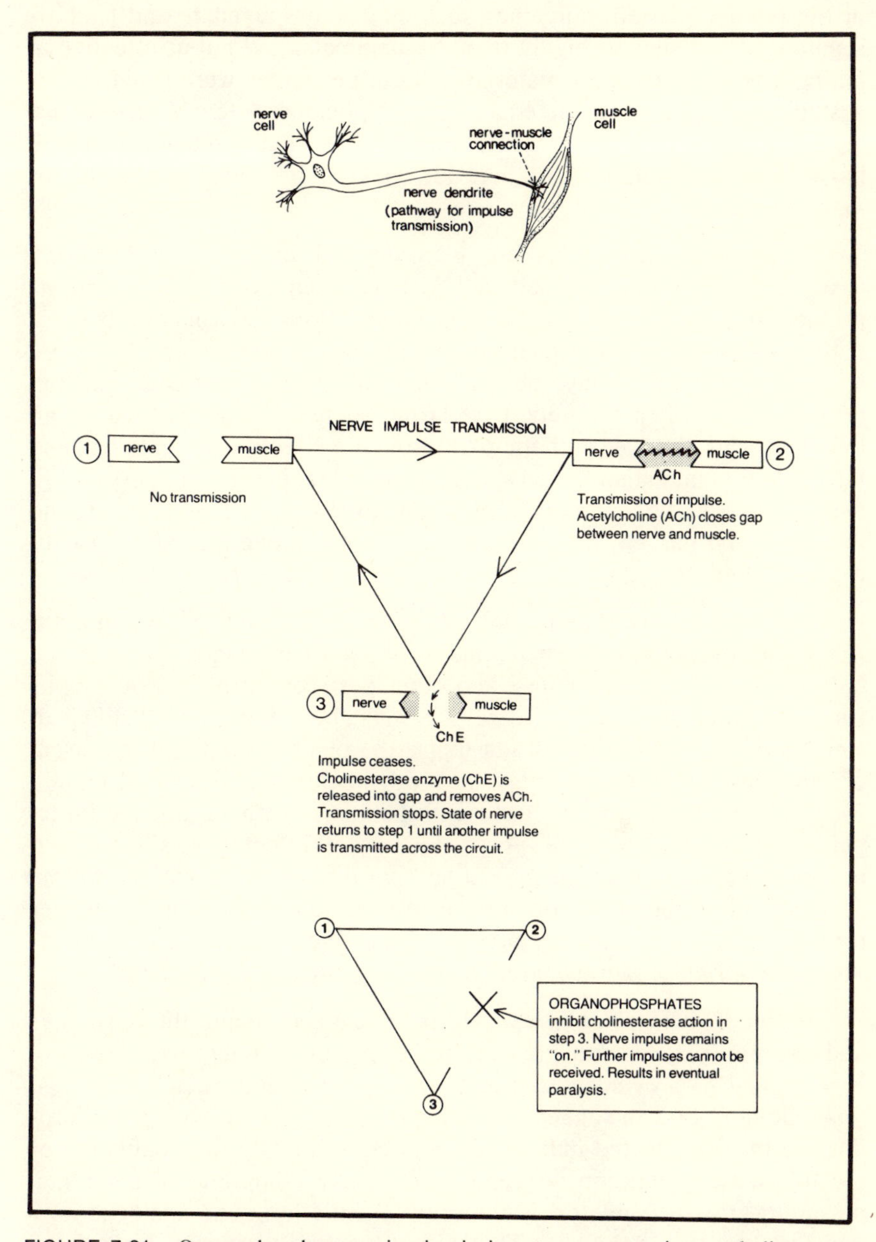

FIGURE 7-21. Organophosphorus poisoning in insects or man owing to cholinesterase inhibition. Top, nerve–muscle connection. Center, normal nerve impulse transmission. Bottom, action of a cholinesterase-inhibiting organophosphorus insecticide on nerve impulse transmission.

of the popularly used inorganics such as calcium arsenate and lead arsenate were persistent, highly toxic to mammals, and not as effective at killing insects as the new materials. These pesticides were generally ingested by the insects and acted as stomach poisons. A few inorganics are still in use today, some of which present few environmental or health hazards. These include boric acid for cockroach control and sulfur for mite control.

Botanicals. Botanicals are plant-derived insecticides. They are toxic substances that are believed to have been evolved by plants as protection from phytophagous insects. Concoctions and leachates of these natural substances have been used for centuries as insecticides. Many are still used effectively today, including nicotine, ryania, rotenone, and pyrethrum. Recently a synthetic pyrethroid has been developed. Although the pyrethrins and the synthetic pyrethroid are effective insect killers and have a low mammalian toxicity, a major problem is their high toxicity to fish and to aquatic arthropods important to fisheries. However, around the home and garden, pyrethrins are probably among the safest insecticides available.

Oils. A variety of petroleum oils have for years played an important role in insect and mite control. Oils have been particularly useful in the control of scale insects, mites, and other pests on citrus and deciduous fruit, nut, and ornamental trees. Oils must be refined and distilled to weights or fractions that will not injure the plant but will kill the insect or mite pest. Heavier oils may be used on dormant trees and certain species of trees even in the growing season. A synthetic organic insecticide may be added to an oil spray to increase its effectiveness. Integrated pest management programs in fruit and nut tree crops frequently rely on dormant oil applications for control because oil applications in winter provide effective control and are less disruptive of natural enemy activities than spring or summer treatments with more toxic materials.

Insect Growth Regulators. These chemicals mimic the hormones and enzymes that regulate growth and development in insects. They may be designed to prevent pest insects from molting into their adult stage (juvenile hormone analogues). In doing this they effectively block reproduction by the affected individuals, and consequently the population of the following generation is greatly reduced. Methoprene (Altosid®), a mosquito growth regulator, is the only juvenile hormone analogue now registered for use in the U.S. The second type of insect growth regulator now available is the chitin synthesis inhibitor diflubenzuron (Dimilin®). Chitin is a major component of the exoskeleton of most insects and other arthropods. Thus, a properly timed application of diflubenzuron will inhibit the formation of a new exoskeleton when an immature insect molts and thereby kills the insect.

Although the specificity of many of these growth regulators and inhibitors is sufficient to make them highly desirable pesticides for use in IPM programs, there are still some problems to be worked out before they can be effectively incorporated as controls in most pest management situations. One problem is the need to precisely time applications of juvenile hormone analogues according to the life cycle of the insects. Another is the lack of assessment of the potential chronic health and ecological problems that may be associated with these new materials, including the scope of their specificity.

Microbial Insecticides. These insecticides utilize insect disease-causing microorganisms or their by-products to control insects and thus are not truly "chemical" insecticides. One great advantage of microbial pesticides is in the specificity of their toxic action and their safety to nontarget species. In some cases, another advantage is their self-multiplicative capacity after being applied. So far, no adverse effects on humans or other vertebrates have been demonstrated for the microbial insecticides that are now commercially available. These include two widely used bacterial pathogens—*Bacillus thuringiensis* ("B.t."), which kills a wide range of caterpillars, and *Bacillus popilliae* (milky spore disease), used to control Japanese beetles—and three viruses. The three viruses are a nuclear polyhedrosis virus toxic to members of the *Heliothis* genus, including the cotton bollworm and tobacco budworm, and polyhedrosis viruses effective against two forest pests, the Douglas fir tussock moth and the gypsy moth. Of the microbial insecticides, *Bacillus thuringiensis* is by far the most widely used. Two and one-half million acres in California alone were treated with this pesticide in 1978. It is registered for use in a number of crops, including vegetables, flowers, cotton, alfalfa, grapes, oranges, and walnuts, as well as on forest and ornamental shade trees. Registration covers use against a wide variety of pests in the order Lepidoptera including the cabbage looper, imported cabbageworm, tobacco budworm, grape leaffolder, omnivorous leafroller, fruit tree leafroller, tomato hornworm, and the California oakmoth. At times growers have used "homemade" microbial insecticides to kill such pests as the alfalfa butterfly, velvet bean caterpillar, and cabbage looper. These sprays are water suspensions of the macerated cadavers of infected caterpillars collected in the field.

In integrated pest management programs where the goal is to achieve effective, selective control with minimum environmental disruption and maximum safety to humans, microbial insecticides are ideal pest control tools. There are many naturally occurring insect pathogens, yet only a few have been thoroughly tested as potential insecticides; undoubtably many candidates with commercial value remain untested or undiscovered.

Pesticides Applied for Control of Plant Pathogens

Since most plant pathogens grow within the tissues of the host plant, are not usually accessible to pesticide application, and reproduce extremely rapidly, these organisms are much harder to control with chemicals than insects and mites. As a result, most of the present use of chemicals for control of fungal and bacterial pests focuses on preventing infection rather than reducing currently existing levels of infestation, making the concept of control action threshold density much less applicable. Most fungicides act as protectants, preventing spore germination and fungal penetration into plant tissues. They are applied to leaves of growing plants, seeds, planting stock, soil, or harvested produce. Eradicant fungicides are usually applied during the dormant or overwintering stages of both crop plant and pathogen. Soil fumigants, primarily methyl bromide, chloropicrin, and carbon bisulfide, may be used to control soilborn pathogens such as *Verticillium* and *Armillaria*. A few pesticides known as chemotherapeutants are used to help control diseases with some success after symptoms appear. By far the greatest number of chemicals available to control plant disease are fungicides. A few bactericides are available, but no chemicals for virus disease control are on the market.

Although used sporadically prior to 1880, both copper and sulfur came into widespread use as broad spectrum fungicides after the outbreak of powdery mildew in the vineyards of France in the last half of the nineteenth century. It was the introduction of these two materials in the form of Bordeaux mixture (copper sulfate, lime and water) for control of that devastating pest that saved the French vineyards from an untimely demise. Bordeaux mixture and various other copper and sulfur compounds are still widely used as fungicides, especially to control mildews. Both copper and sulfur, however, produce phytotoxic reactions in some crop plants, and for this reason the synthetic organic fungicides are growing in popularity.

The diothiocarbamates are the most widely used group of synthetic organic fungicides. They are also sulfur-containing and include thiram, maneb, ferbam, ziram, Vapam®, and zineb. Other popular organic fungicides include pentachlorophenol (PCP), used extensively as a wood preservative; pentachloronitrobenzene (PCNB), a persistent soil fungicide; dinocap for powdery mildews; and chlorothalonil, captan, folpet, and captafol.

Systemic fungicides are taken up by the plant roots and travel throughout the plant to its growing points. In this way they may offer "therapeutic" treatment to already infected plants. Only a few are commercially available. These include carboxin, oxycarboxin, chloroneb, and benomyl.

Antibiotics are substances produced by one organism and toxic to another organism, usually a microorganism. One of the most commonly known antibiotics is penicillin, a fungus-produced toxin used against bacterial diseases in people and other animals. Antibiotics used for the control of plant pathogens are usually systemic, translocated from one part of the plant to another. Streptomycin, used in the control of bacterial diseases of plants, is the most familiar of these.

Herbicides

Materials manufactured for the control of weeds and other unwanted vegetation now account for more than one-half of the pesticides used and produced in this country. Herbicides may be broadly classed as selective or nonselective. Selective herbicides kill or stunt weed growth without harming the crop plants beyond the point of adequate recovery. Nonselective herbicides kill all plants and must be used before a new crop is planted or used in areas where no plant growth is desired. Paraquat and amitrole are examples of nonselective herbicides. Most selective herbicides become nonselective at higher rates or when improperly applied.

Herbicides may kill only the plant parts on which the material is applied (contact herbicides) or they move within the plant, destroying roots and other parts not exposed to the pesticide (translocated herbicides). The phenoxy herbicides, including 2,4-D and 2,4,5-T, silvex, and MCPA, are the best-known translocated herbicides.

The phenoxy herbicides are sometimes called the "hormone" herbicides because they act in a hormonelike manner as growth stimulants, causing cell division and other growth activity to happen so rapidly and relentlessly that the plant literally grows itself to death. They affect only broad-leaved plants, so they are used in situations where grass, grains, conifers, and other narrow-leaved species must be grown without broadleaf competition. Other well-known translocated herbicides include dalapon, dicamba, and picloram. Paraquat, diquat, dinoseb, propanil, bromoxynil, and weed oils are commonly used contact herbicides.

Soil-applied herbicides incorporated into the soil before or soon after planting may be translocated from the roots of the growing seedling to the shoot. Others act directly on the root. Many herbicides are soil-applied, including atrazine, benefin, bromacil, chloramben, molinate, monuron, nitralin, prometryne, propham, cycloate, diuron, simazine, endothall, EPTC, and trifluralin. A few materials like dinoseb, dicamba, and picloram may be soil- or foliage-applied. Soil fumigants such as methyl bromide and carbon bisulfide act as nonselective soil herbicides.

Pesticide Formulations

Pesticides are applied as fumigants, aerosols, sprays, smokes, dusts, granular pellets, residual fumigants, baits, and seed or furrow treatments. They are applied in the soil, above the soil, onto plant leaves or other surfaces, and directly to animals or their housing. They may be impregnated into cloth, lumber, or paper, and may be taken up or eaten and translocated systematically in the tissues of plants or animals or to the feces of animals. They may be formulated together—e.g., two different insecticides, a fungicide and an insecticide—or with a fertilizer. They may contain synergists or activators that enhance the toxic properties of the pesticide. They may be formulated with sticker–spreader additives that facilitate distribution of the pesticide over the leaf and increase its persistence in the environment. Table 7-6 shows some of the more common formulations of pesticides. Choice of formulation is very important and is dependent on both pest and crop biology and environmental conditions. Formulations such as baits can limit environmental contamination in many situations as compared to dusts and sprays; however, the safer pesticide formulations are not always effective. Each possible usage must be carefully studied.

The case of encapsulated methyl parathion illustrates how a new formulation developed for its apparent increased safety can cause unexpected environmental problems. Formulation of the pesticide in tiny beads of plastic allowed the slow release of the toxic material, thereby increasing its life in the environment and diminishing its hazard to human handlers. By using these incapsulated materials growers were able to reduce the number of applications, be less precise about application times, and reduce worker injuries. Unfortunately, the microencapsulated beads were the same size as pollen grains, and bees carried this highly toxic material back to their hives. The increased life of the material in the field also prolonged the period of hazard to bees. As a result, after the introduction of encapsulated methyl parathion, beekeepers began to experience some of the worst losses in their industry in recent memory.

Pesticide Toxicity

Pesticides are, by definition, poisons. They are commonly toxic to a wide variety of organisms in addition to the target pest. It is essential

TABLE 7-6
Common Formulations of Pesticides[a,b]

1. Sprays (insecticides, herbicides, fungicides)
   a. Emulsible concentrates
   b. Water-miscible liquids
   c. Wettable powders
   d. Flowable or sprayable suspensions of ground toxicants in water
   e. Water-soluble powders
   f. Oil solutions
   g. Ultra-low-volume concentrates
2. Dusts (insecticides, fungicides)
   a. Undiluted toxic agents
   b. Toxic agents with active diluent, e.g., sulfur
   c. Toxic agents with inert diluent, e.g., pyrophyllite
   d. Aerosol "dust"—silica in liquefied gas propellant
3. Granulars (insecticides, herbicides)
4. Aerosols (insecticides)
5. Fumigants (insecticides, nematicides)
   a. Space and stored products treatment
   b. Plastic strips impregnated with volatile insecticide
   c. Soil treatment liquids that vaporize
6. Impregnating materials (insecticides, fungicides)
   a. Wood preservatives
   b. Mothproofing preparations for woolens
7. Impregnated shelf papers, strips, tapes (insecticides)
8. Baits for grasshoppers, crickets, ants, slugs
9. Animal systemics (insecticides, parasiticides)
10. Animal dressings (insecticides)
11. Fertilizer–insecticide combinations
12. Encapsulated insecticides
13. Insect repellants
    a. Aersols
    b. Rub-ons

[a] This list is incomplete, containing only the more common formulations.
[b] From Ware (1975).

that the pest manager know the possible impact of pesticide application on all nontarget organisms.

In animals, pesticides may show oral toxicity (if eaten or drunk), dermal toxicity (if absorbed through the skin), subcutaneous toxicity (if injected just below the skin), intramuscular toxicity (if injected into the muscles), intravenous toxicity (if injected into a vein), intraperitoneal toxicity (if injected into the viscera), or inhalation toxicity (if inhaled or breathed in). Toxicity symptoms may manifest themselves soon after exposure (acute poisoning) or weeks to years after exposure (chronic poisoning). Some exposures will result in both types of poisoning symptoms.

The acute toxicity of a pesticide is generally indicated by a measurement called the LD_{50}. The LD_{50} describes the *median lethal dose* to test animals, that is, the dose that kills exactly 50% of the animals exposed to it under the prescribed experimental conditions (Figure 7-22). LD_{50} is expressed in a ratio of milligrams (mg) of the toxic substance per kilogram (kg) of animal body weight. LD_{50} tests are generally carried out on rats, although mice, rabbits, monkeys, cats, dogs, and guinea pigs may also be tested. Choice of animal is important because many times different mammal species will show different pesticide tolerances. LD_{50}s generally vary between the sexes as well. Knowing this, of course, it is often hard to extrapolate precisely what an LD_{50} means in terms of the pesticide's toxicity to man. LD_{50} measurements may also be used to indicate the toxicity of a pesticide to the targeted pest species.

Many pesticides also have teratogenic, mutagenic, or carcinogenic properties. Currently, testing methods for these adverse effects are difficult to interpret and, until the recent development of bacterial culture tests (e.g., the Ames test) and tissue culture tests, were extremely laborious. Pesticides are often widely used for a long period of time before these properties are known. Because not all pesticides have been adequately examined, it is likely that other materials currently in use may be shown to be carcinogenic, mutagenic, or teratogenic at a future date.

Because of the highly toxic nature of pesticides, the pest manager must take all possible steps to insure minimum risk to wildlife, people, and livestock exposed to these poisons. Some suggested precautions are listed in Table 7.7.

Selectivity

Because of the potential adverse impacts of pesticide use, researchers have been trying to minimize these undesirable side effects. While one way to minimize adverse impacts of pesticides is to substitute other non-pesticide methods of control, another approach that has received much attention and support is the development of pesticide chemicals or application techniques that pose the least risk potential to nontarget organisms. In other words, pesticides are used in ways which selectively kill unwanted organisms, leaving most other plants and animals unharmed.

The selective use of chemicals can be carried out in a number of ways. The first involves the choice of materials. In the past, the development of pesticides possessing "broad-spectrum" toxicity, that is, the ability to kill a wide range of pests, was unquestionably favored over the development of more narrowly toxic materials. A broad-spectrum pesticide would have a larger market and could be produced in larger, more profitable quantities than a selective pesticide with the capacity for killing

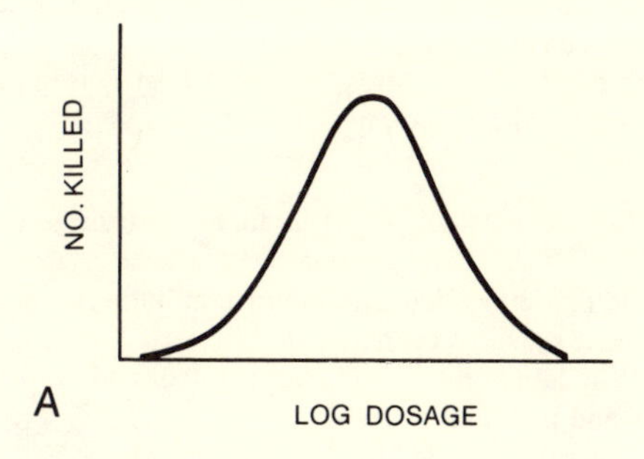

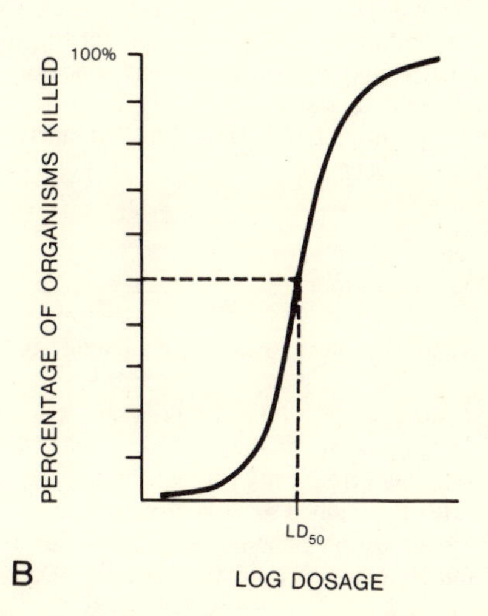

FIGURE 7-22. Determination of LD_{50} or median lethal dose. (A) shows that the susceptibility of individuals to toxic substances varies among the population of an animal species. Thus, some are killed by a small dose, a large number are killed by a somewhat larger dose, and a few can tolerate even more of the toxic substance. (B) shows how this information is converted into an LD_{50} determination. The percentage killed includes the individuals killed by smaller dosages than that indicated [thus the graph is cumulative and of a different shape than (A)].

TABLE 7-7
Suggestions for Safe Handling of Pesticides[a]

Basics for the employer

1. Inform all employees of the hazards associated with pesticide use.
2. Have prearranged medical services available.
3. Post the telephone number and address of the nearest Poison Control Center at every telephone in the work situation.

Selection of pesticides

1. Identify the pest to be controlled and, if in doubt, consult your County Agent or other authority.
2. Select the pesticide recommended by a competent authority and consider pesticide residue effcts on next season's crops.
3. Make certain that the label on the container is intact and up-to-date; it should include directions and precautions.
4. Purchase only the quantity needed for the current season.

Transporting pesticides

1. Hazardous materials should be conspicuously marked on the container.
2. Open-type vehicles are preferred to closed vehicles for transportation of volatile pesticides and those that would give off poisonous or noxious fumes if their containers were accidentally unsealed.
3. All containers should be tightly closed. Do not haul ruptured bags, cans, or drums.
4. Always be prepared for an *accident*.

Storing pesticides

1. Lock all pesticide rooms, cabinets, or sheds.
2. Do not store pesticides where food, feed, seed, or water can become contaminated.
3. Store in a dry, well-ventilated place, away from sunlight, and at temperatures above freezing.
4. Mark all entrances to storage PESTICIDES STORED HERE—KEEP OUT.
5. Keep pesticides in original containers only—closed tightly and labeled.
6. Examine pesticide containers frequently for leaks and tears. Dispose of leaking and torn containers, and clean up spilled or leaked material immediately.
7. Where possible, a sink for washing should be located in or near storage areas.
8. Keep an inventory and eliminate all outdated materials. Date containers when purchased.
9. Take precautions for potential fire hazards.

Handling and mixing pesticides

1. Before mixing, carefully read the label directions and current official state recommendations of the Cooperative Extension Service.
2. Wear appropriate protective clothing and equipment as specified on the label.
3. Handle pesticides in a well-ventilated area. Avoid dusts and splashing when opening containers or pouring into the spray apparatus. Do not use or mix pesticides on windy days.
4. Measure the quantity of pesticide required accurately, using the proper equipment.
5. Do not mix pesticides in areas where there is a chance that spills or overflows could get into any water supply.
6. Clean up spills immediately. Wash pesticides off skin promptly with plenty of soap and water. Change clothes immediately if they become contaminated.

(Continued)

TABLE 7-7 (*Continued*)

Applying pesticides

1. Wear the appropriate protective clothing and equipment as required for toxic materials.
2. Make certain equipment is calibrated correctly and is in satisfactory working condition.
3. Apply only at the recommended rate. To minimize drift, apply only on a calm day and do not work through clouds or drift of unsettled dusts or sprays. Do not contaminate livestock, feed, food, or water supplies.
4. Avoid damage to beneficial and pollinating insects by not spraying during periods when such insects are actively working in the spray area. Notify neighboring beekeepers, as required by legislative regulations, at least 24 hours before application so precautionary measures can be taken.
5. Keep pesticides out of mouth, eyes, and nose. Do not use mouth to blow out clogged lines or nozzles.
6. Observe precisely the waiting periods specified between pesticide application and harvest or reentry time. Keep people and animals out of treated area as indicated on label.
7. Clean all equipment used in mixing and applying pesticides according to recommendations. Do not use herbicide application equipment for applying insecticides.
8. After handling pesticides, bathe skin throughly with soap and water and change clothing.
9. Keep complete and accurate records of the use of pesticides.
10. If symptoms of poisoning occur during or shortly after the use of or exposure to a pesticide, call the physician or take the patient to the hospital immediately. Take the pesticide label with you. It is wise to alert your own family members and the family physician before using highly toxic pesticides.

Disposing of empty containers and unused pesticides

1. "Empty" containers are never completely empty.
2. Don't reuse the container.
3. Empty the contents and bury unused chemicals at least 18 inches deep in an isolated (marked) location away from water supplies.
4. Break glass containers and bury the pieces as (3) above.
5. Burn empty fiber and paper containers, except herbicide containers. Stay well away from the smoke. Burn in designated areas only.
6. Rinse metal containers twice with water, punch holes in top and bottom, and bury containers and rinses as in (3) above.
7. If a metal container cannot be rinsed, punch holes in top and bottom and bury as in (3) above.

[a] From Ware (1975).

only a few pest species. However, these broad-spectrum pesticides, especially the broad-spectrum insecticides, have had disastrous results on natural enemies, honeybees, and many other beneficial and nontarget species, including wildlife, livestock, and humans. Entomologists are now increasingly recognizing the need to protect not only humans and livestock but also the natural enemies and pollinators in managed ecosystems. Their concern will, it is hoped, convince pesticide manufacturers to develop more of these selective materials. The microbial insecticides are promising candidates. Figure 7-23 shows the selective action of two insecticides on mite populations in an apple orchard.

Until adequate numbers of truly selective pesticides are developed, the integrated pest manager must rely on ecologically selective methods to harmonize (i.e., integrate) necessary pesticide treatments with other control tactics and natural mortality factors. The preprogrammed, unilateral application of broad-spectrum poisons onto broad expanses of land (whether there is a need or not) requires none of the ecological knowledge necessary for selective use of pesticides. But in integrated pest management, the use of pesticides, like other IPM tools, must be approached with a good ecological understanding of the pest problem and the consequences of different sorts of solutions. Some of the most common ecologically selective methods of pesticide use are described below.

Timing of Application. For their most effective use, pesticides, like

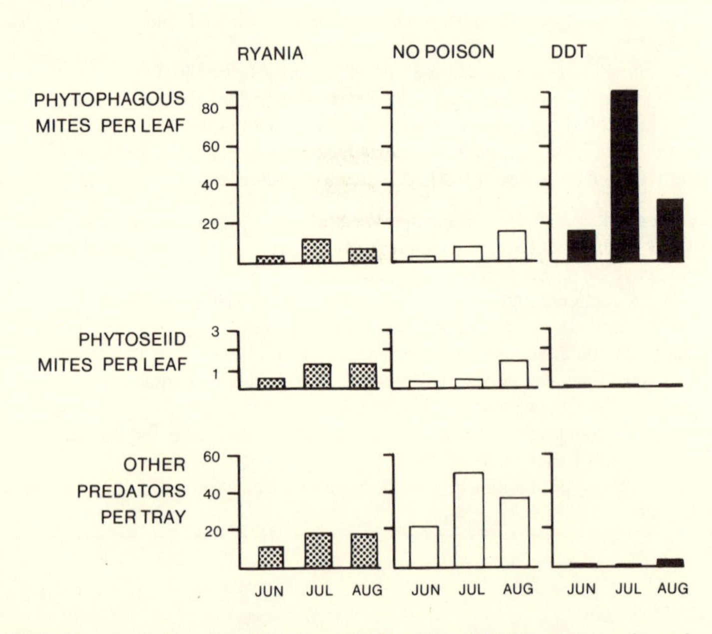

FIGURE 7-23. Pesticide selectivity in an apple orchard. DDT was very damaging to phytoseiid and other predator mite populations and thus increased population levels of the phytophagous pest mites. On the other hand, ryania had little effect on phytoseiid mites and a less damaging effect on populations of other predators. Thus ryania might be a good selective insecticide to use against a nonmite pest that is not being adequately controlled by natural mortality factors (after MacPhee and MacLellan, 1971).

cultural controls, must be aimed at the "weak point" in the pest's life cycle. Many insecticides are effective only on the early active stages of the pest. Thus, applications when pests are in egg, late larval, pupal, or adult stages may be largely ineffective. Proper timing of pesticide application requires sampling of population numbers to ascertain the age composition of the pest populations. Light and pheromone traps, which catch only adult insects, have been invaluable in the proper timing of pesticide applications for pests which occur as brooded populations. With a knowledge of when peak adult populations occur, the pest manager can effectively estimate when the most vulnerable stage of the next generation will occur, and if these do approach dangerous levels, he should apply pesticides appropriately. In many agricultural crops, this procedure has reduced the number of annual applications drastically.

Pesticide applications should also be timed to avoid drift (e.g., on windless days, in early morning or evening) and undue damage to nontarget organisms within the treated areas (e.g., avoiding sprays at times of the day when bees and other pollinators are in the areas).

Placement of Pesticides. The confinement of pesticide application to a very localized area can greatly minimize the amount of poisons in the managed environment. Seed treatment for insect, plant pathogen, and nematode control is an example. Preplant application to seeds requires only a fraction of the pesticide that would be needed for a soil application.

Another way to minimize amounts of pesticides used for insect and other animal control is to attract pests to a trap crop or a trap (baited with food, pheromones, lights, or some other attractant) and poison the pests there. Traps can often be designed to minimize leakage of the toxic substance into the environment.

Spot treatments are also a valuable way of reducing pesticide use and avoiding ecological disruption. Often pests will concentrate only in certain parts of the managed area. These localized spots can be delineated and treated, allowing the survival of beneficial and other nontarget organisms in other parts of the field, orchard, or forest.

Application Technique and Equipment

Much pesticide is wasted during any application operation; aerial spray applications are among the most wasteful. As Figure 7-24 shows, less than 1% of an aerial spray insecticide application can be expected to hit the target—i.e., get inside the pest insect. Applications of materials formulated as dusts most frequently result in drift of the pesticide outside the target area. Equipment and techniques that minimize this waste are necessary to provide more selective pesticide use. Especially important

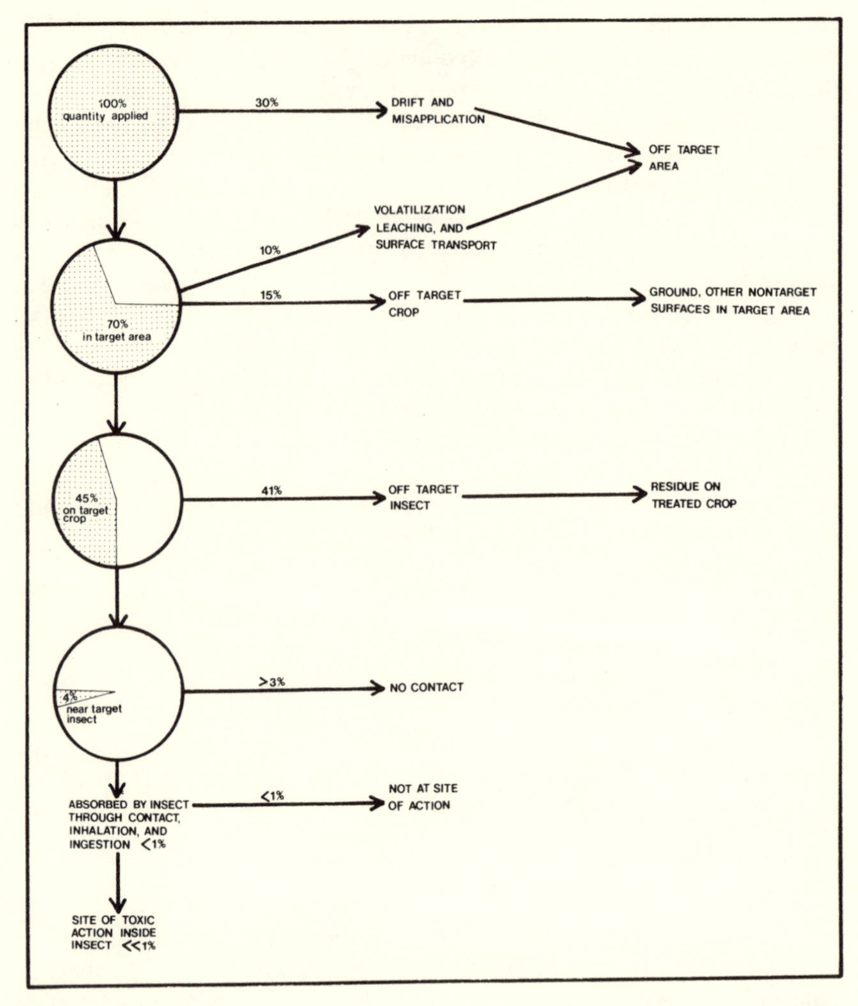

FIGURE 7-24. Aerial foliar insecticide application: typical losses between spray nozzle and site of toxic action (after von Rumker *et al.*, 1974).

are careful calibration of application equipment nozzles to produce more uniform and larger droplets, choice of pesticide formulations least subject to drift losses, better monitoring and use of meteorological data for determining best conditions prior to application, and making sure applicators are adequately trained to carry out the responsibilities involved in applying pesticides.

Future advances in application equipment and technique may assist in the selective placement of pesticides. For instance, many pests only inhabit certain areas of the host plant. If economically practical methods and equipment could be designed that selectively treated only these areas, much current waste could be eliminated. For example, it has been found that in California peaches, the oriental fruit moth occurs primarily in the upper one-third of the trees. Consequently, in an integrated pest management program for peaches, sprays for this pest are directed to the top third of the trees. This greatly reduces insecticide costs and preserves natural enemies in much of the tree.

Pesticide use should always be approached with caution. As the most disruptive control action option, it is the last choice of the pest manager in an integrated pest management program and should always be justified by careful sampling and with a consideration of the natural control factors operating in the ecosystem.

As has been shown in previous chapters, the use of an insecticide for one pest problem often creates new problems far worse than those caused by the original "controlled" pest. Usually the eruption of a new pest problem cannot be economically justified by the control of the initial pest.

For the most environmentally sound protection from pests, integrated pest management must depend more on the prevention of future pest problems by use of resistant varieties and enhancement of natural enemy activity or modification of the environment to depress pest populations than by controlling already economically damaging populations. This goal demands much research, broad knowledge and foresight on the part of the pest manager and thoughtful use of control tactics.

Although this chapter has presented the different types of control action options singly, their incorporation into an IPM program must always be integrated with other management actions and particularly with regard to naturally occurring mortality factors (biological and physical) operating in the managed environment. IPM depends heavily on the "check and balance" system. If a pest organism has many limiting factors operating on its population growth, it is less likely to suddenly increase to damaging levels. If these limiting factors are permanent or long-term influences in the ecosystem, such as natural enemies or environmental modifications, the "check and balance" system has an even stronger effect.

The following chapter will discuss successful IPM programs in several crops and resource ecosystems and illustrate how several control action options can be integrated for economically and environmentally sound pest control.

CHAPTER 8

Case Histories in Integrated Pest Management

In the previous pages a wide variety of IPM cases have been discussed. Those perhaps represent the majority of programs in the USA. In this chapter, a few case histories of IPM implementation are described in greater detail.

ALFALFA IN CALIFORNIA

The California alfalfa agro-ecosystem is an appropriate place to start a survey of IPM case histories. Alfalfa was one of the first crops in California in which a holistic approach to pest managenent was instigated, and it provided major impetus for the coming of age of the integrated control concept.

Irrigated alfalfa covers more than 1.5 million acres of cropland in California. Two-thirds of this area is concentrated in California's great Central Valley; the rest is scattered in various places, ranging from the low desert valleys bordering Mexico to the transmontane valleys of Northeastern California and the valleys along the Pacific coast. As might be expected, each area has a different climatic regime and a slightly different set of pest problems. The present discussion will concentrate on the situation that characterizes the bulk of this alfalfa acreage—that growing in the Central Valley.

Alfalfa is an important crop, not only for the several hundred million dollars it brings annually to the California alfalfa growers themselves, but also because so many other industries are dependent on alfalfa production.

It is a major feed source for livestock and poultry. Besides the use of fresh alfalfa in California, tons of the crop are sold and shipped as pellets or cubes to other areas. Alfalfa is also an important rotation plant, and it enhances soil quality by adding nitrogen and organic matter, increasing water filtration rates, and improving the soil structure.

The first step toward an integrated pest management program in California alfalfa was the development of a "supervised" control approach to the management of the alfalfa caterpillar (*Colias eurytheme*). The early workers on this project, Michelbacher and Smith, observed that natural enemies of this pest, including a parasitic wasp (*Apanteles medicaginis*) and a nuclear polyhydrosis virus, were often able to rapidly reduce populations nearing economically threatening levels.

After recognizing the extent of naturally occurring biological control, Michelbacher and Smith introduced a pest control strategy for the alfalfa caterpillar that stressed the importance of avoiding insecticide applications when natural enemy populations were high enough to control the pest. To implement this recommendation they developed an economic threshold of 10 healthy, nearly mature alfalfa caterpillar larvae per sweep. The key that distinguishes this as an integrated program is the modifier "healthy." Michelbacher and Smith recognized that if substantial numbers of collected larvae were not healthy (that is, were parasitized by the wasp or were dying from the virus disease) that natural control factors were catching up with the caterpillar population and would probably be able to keep the pest below economically damaging levels.

The need for skilled entomologists (or pest management specialists–consultants) to carry out monitoring programs and assist growers in making control action decisions was recognized at this time. Only a specially trained person could correctly identify pests and their state of health. The first supervised control expert in California, Kenneth S. Hagen, was hired in 1946 to implement the integrated control program against the alfalfa caterpillar.

The arrival of the devastating spotted alfalfa aphid (*Therioaphis trifolii*) in California in 1954 provided the impetus for further development of the integrated control approach. This aphid spread more rapidly and created more havoc than any other invading pest in memory. In 1955 losses attributed to it in California were over $300,000. In 1956 this figure rose to $9 million and in 1957 to $10.6 million. The pest threatened to wipe out the alfalfa industry in California.

However, the widespread implementation of an integrated control program in 1958 dropped losses to $1.7 million in just one year. With the addition of a resistant variety of alfalfa into the system in 1960, the integrated pest management program for the spotted alfalfa aphid was complete, and in California the pest now rarely assumes injurious status.

The first step in the integrated pest management program against the spotted alfalfa aphid was to try to understand the ecological factors that were enhancing the pest's activities. A survey of the naturally occurring biological control of the aphid revealed that lady beetles (*Hippodamia* spp.) were such effective predators in the spring months that there were no aphids left for the next generation of lady beetles (in the early summer months) to feed on (Figure 8-1). Most would starve or be forced to migrate, and spotted alfalfa aphid populations would then rise. Then, just as the lady beetle predators (again having enough to eat) began to move in again

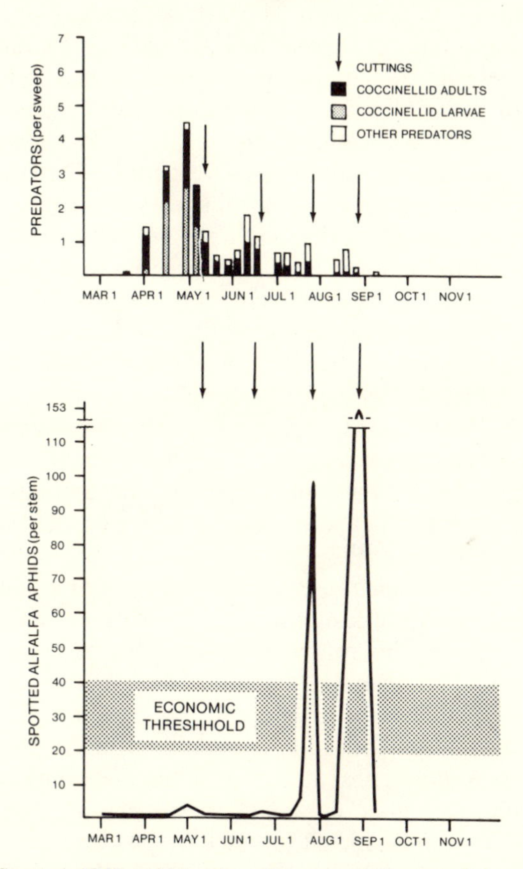

FIGURE 8-1. Spotted alfalfa aphid and predator population trends in an unsprayed alfalfa field near Cressey, Merced County, 1956. Stippled area represents the economic threshold for spotted alfalfa aphid. Predators include coccinellid adults (■), coccinellid larvae (▒), and other potential aphid predators (□). (These data were collected before the introduction of parasites into the area.) Note how predator populations decrease in late spring (when aphid populations are too low to maintain large predator numbers). The dated vertical guidelines indicate cutting times; these show how a cultural practice (cutting the alfalfa) can be timed to lower aphid populations (from Smith and Hagen, 1959).

or increase, the general practice was to apply an insecticide spray, decimating the predator populations; and a more rapid resurgence of the aphid would result. Secondary pest outbreaks of insects such as the alfalfa caterpillar and beet armyworm, whose natural enemies were also killed, would require additional insecticide treatments. The eventual development of insecticide-resistant strains of the spotted alfalfa aphid completed the symptoms of the "pesticide treadmill."

The integrated pest management program was instigated to get alfalfa growers off this pesticide treadmill. Several tactics were involved. First, classic biological control was undertaken; three species of hymenopterous parasites were imported from the Near East and Europe and established in California. These introduced natural enemies contributed significantly to the control of the aphid. Second, a cultural control, strip harvesting of alfalfa (as discussed in Chapter 6, Figures 6-5 and 6-6), was employed. The practice of strip harvesting enhanced the action of both the introduced and naturally occurring biological control agents by providing refuges for these natural enemies. In another cultural control wrinkle, timely irrigation was utilized to enhance activity of a virulent fungus that caused a disease in the aphid. Third, a sound economic threshold for the pest was established, and a sweep-net sampling method for predators was standardized. And fourth, a selective insecticide, demeton (Systox®), was developed that killed the spotted alfalfa aphid but was less destructive to natural enemy populations. This material provided an effective yet relatively nondisruptive control when sampling indicated that chemical treatment was the only action option. The need for any insecticide treatments for the spotted alfalfa aphid was eliminated by the introduction of a fifth tactic, the use of alfalfa varieties resistant to the spotted alfalfa aphid. Although the chemical control facet has been eliminated by the use of resistant alfalfa varieties, biological controls and, to some degree, cultural controls are still important limiting factors on the spotted alfalfa aphid population. Consequently, the control of this pest remains truly "integrated."

To assure the success of the integrated control program against the spotted alfalfa aphid it was essential to also eliminate unnecessary insecticide applications for other pests in alfalfa as well. Thus, economic thresholds were developed for all the major pests (see Chapter 7, Table 7-2), and a cultural program of early mowing for the alfalfa weevil was able to keep this pest under control without chemical applications in many instances. Use of microbial insecticides for control of loopers and the alfalfa caterpillar has enhanced the program in recent years. The need for a minimum-pesticide program in alfalfa has been further emphasized with the growing awareness of the dangers of toxic residues in feed fed to livestock, especially to dairy cattle. More recently, however, two new insect pests, the Egyptian alfalfa weevil and the blue alfalfa aphid, have been posing new problems, and efforts to integrate their control into the program are underway.

INTEGRATED PEST MANAGEMENT IN APPLES IN NOVA SCOTIA

Apples are grown all across the North American continent. Pest complexes vary from region to region, and pest management programs must necessarily reflect these regional differences. This case history presents the "grand-daddy" of all integrated pest management programs in deciduous fruit tree crops, the Nova Scotia apple program, pioneered by A. D. Pickett, F. T. Lord, and colleagues.

Although the beginnings of the Nova Scotia IPM program predate the DDT era, the development of this program was spurred on by the same type of "pesticide treadmill" syndrome that characterized the breakdown of standard chemical control programs in other crops in later decades. Because of the high value and the relatively low tolerance for wormy apples at the marketplace, orchards were among the few agricultural ecosystems to receive regular, heavy pesticide applications before the 1940s; this explains the early appearance of the pesticide treatmill syndrome in many orchards.

Frequent spraying of orchards with oils and various types of inorganic pesticides was a common practice from the turn of the century. Accordingly, the first documented case of insecticide resistance involved an orchard pest (a strain of the San Jose scale in 1914 resistant to lime–sulfur sprays); and all five pre-1940 reports of insecticide-resistant pest strains were orchard pests, including a strain of the codling moth (a key apple pest) reported resistant to arsenical sprays in 1928.

In Nova Scotia apple orchards in the 1920s and 1930s, more frequent and heavier dosages of insecticidal chemicals were required to suppress pests as time went on. Several, including the codling moth (*Laspeyresia pomonella*) and the eye-spotted bud moth (*Spilonota ocellana*), became more frequent and increasingly destructive (i.e., target pest resurgence). Secondary pest outbreaks also plagued the Nova Scotia orchards. For instance, the oystershell scale (*Lepidosaphes ulmi*) became a serious pest for the first time in the 1930s. Two others, the European red mite (*Panonychus ulmi*) and the grey banded leafroller (*Argyrotaenia mariana*) were major pests by the early 1940s.

The Nova Scotia program developed a step at a time. The oystershell scale problem was tackled first. Previous to 1930, naturally occurring biological control in the orchards was sufficient to keep this pest well below economically damaging levels. However, studies in the 1940s showed that predator and parasite populations had been reduced by use of chemicals to levels too low to exert effective biological control on the scale population. The beneficial insect populations were not the nontarget victims of insecticide applications in this case but had been killed off by sulfur-based fungicides applied for plant pathogen control. A simple substitution of copper-based or ferbam fungicides for the sulfur-based materials was instituted, and as a result, parasites and predators of oystershell scale rebounded and were again able to effectively control the pest (Figure 8-2). Continued use of such selective pesticides for plant pathogen control has permitted full biological control of the oystershell scale since the 1940s.

The next secondary outbreak pest to be tackled was the European red mite. In unsprayed orchards these mites were almost always kept well below economic thresholds by an effective predator complex. However, insecticide treatments for codling moth and other key pests had so interfered with the activities of the European red mite's natural enemies that the mite became a major pest by the mid 1940s. The introduction of selective insecticides such as ryania (see Chapter 7, Figure 7-23) for control of the codling moth together with a new philosophy of applying pesticides only when absolutely necessary allowed predator populations to recover, and they effectively eliminated the European red mite problem (Figure 8-3, Table 8-1).

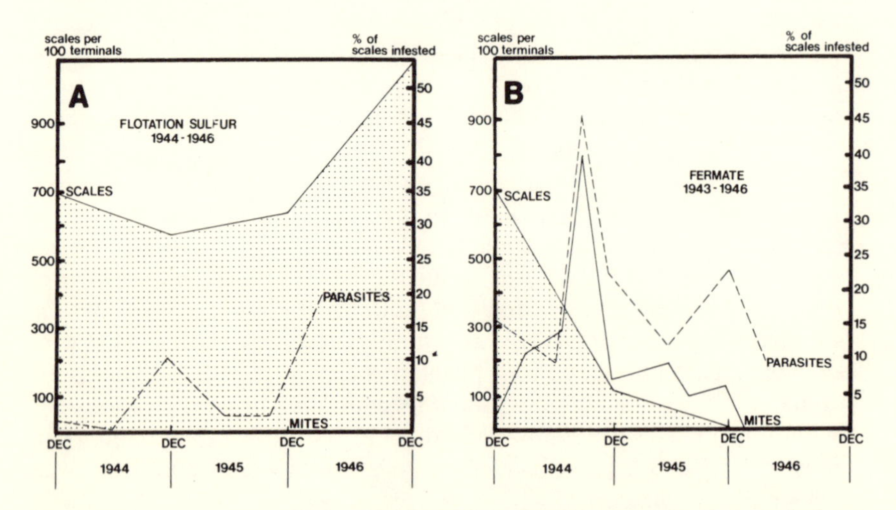

FIGURE 8-2. In the 1940s, use of sulfur-based fungicides for plant pathogen control caused severe outbreaks of the oystershell scale through the destruction of its natural enemies (A). (B) shows how a switchover to a more selective fungicide, ferbam, allowed survival of these beneficials and thus reduced oystershell scale infestations. Natural enemies of this scale include hymenopterous parasites and predatory mites (after Lord, 1947).

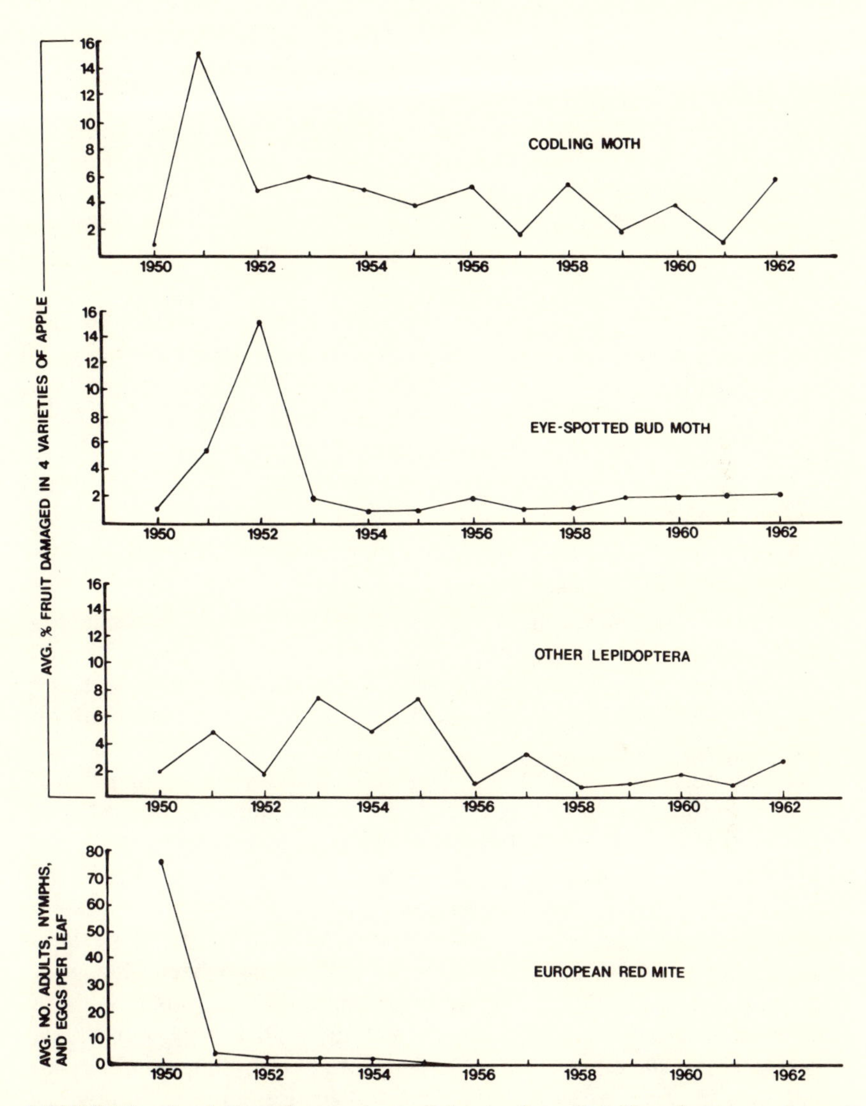

FIGURE 8-3. Populations of pest and potential pest arthropods in Nova Scotia experimental apple orchard plots. Note that reduction of pest population lagged one to two years behind instigation of IPM program in 1950. This lag represents time required for natural enemy populations to recover from effects of pre-1950 heavy pesticide use. Table 8-1 shows pesticides used on these plots, 1950–1962 (data from Patterson, 1966).

TABLE 8-1
Insecticides Applied on Experimental Apple Orchard Plots, IPM Program, and Pests They Were Intended to Control, 1950–1962[a,b]

Year	Insecticide	Rate/acre[c]	Date applied	Pest
1950	Nicotine sulfate, 40%	6 pt	July 15	Eye-spotted bud moth
1951	Nil			
1952	Lead arsenate	10.6 lb	July 11	Forest tent caterpillar
1953	Nicotine sulfate, 40%	5 pt	May 1	Forest tent caterpillar
1954	Nil			
1955	Nil			
1956	DDT, 50w	0.53 lb	June 11	Cankerworm
1957	Nil			
1958	Lead arsenate	10.3 lb	June 5	Cankerworm
1959	Malathion 25w	3 oz	May 29	*Atractotomus mali*
1960	Lead arsenate	11.4 lb	May 20	Cankerworm
1961	Nil			
1962	Lead arsenate		May 31	Winter moth

[a] From Patterson (1966).
[b] Pest populations in these plots are graphed in Figure 8-3.
[c] Each treatment with an insecticide was applied uniformly over the whole orchard.

Similarly, a third secondary outbreak pest, the lecanine scale (*Lecanium coryli*), which became a major pest in the late 1950s, was returned to a nondamaging level after the introduction of selective pesticides and a general reduction in insecticide use that permitted survival of its natural enemies.

A new pest, the winter moth (*Operophtera brumata*), a major pest of oaks and other deciduous trees, was accidentally introduced into North America from Europe in the 1930s and began to appear in Nova Scotia apple orchards at damaging levels in the late 1950s. Two parasites of the moth were successfully introduced from Europe to control the pest in forests, and these together with other natural control factors, especially an introduced virus disease and climatic limitations, have been successful in keeping the pest below economic levels in most deciduous forest and some apple orchard situations (Figures 8-4 and 8-5). But the introduced natural enemies have not been as successful in controlling winter moth populations in orchards as in forests partly because of disruption by codling moth sprays and better synchronization of the moth with the cultivated apple host. However, in orchards near forested areas, the imported natural enemies are often able to keep winter moth populations below economic levels. Otherwise, a selective insecticide is applied.

The apple maggot (*Rhagoletis pomonella*) is a key pest in Nova Scotia apple orchards. Natural enemies are generally unable to keep this pest below economic thresholds, and control depends on an integration of cultural controls (removal of wild hosts, disposal of infested drop-fruit, and general sanitation) and the use of selective insecticides when deemed necessary.

Populations of the codling moth are often kept below economic thresholds by a combination of natural factors including insect parasites and predators, birds, disease, and cold weather. The codling moth has only one generation a year in Nova Scotia and is thus not as serious a pest as it is in Michigan, New York, and other more southern areas where it has a second or third generation each year. However, populations of the codling moth may become damaging, and selective insecticides are applied when sampling indicates that artificial controls are necessary. Thus a maximum of one well-timed application a year is sufficient to control the codling moth in Nova Scotia, and often even this treatment is not necessary.

The integrated pest management program has been so successful that nearly 90% of Nova Scotia apple growers have been using the program continuously since the 1950s. IPM programs for apple pest complexes in other parts of the North American continent have also achieved considerable success. One of these is the Washington IPM program in apples, described in Chapter 3, where maintenance of an alternate host and proper timing of insecticide applications together allow survival of an important pest mite predator, thus increasing effectiveness of biological control and reducing need for pesticide applications (Figure 3-14).

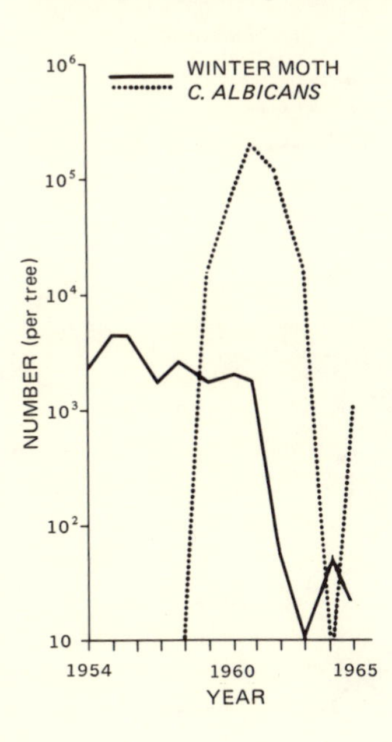

FIGURE 8-4. Winter moth population reduction in a deciduous forest after the introduction of the parasite *Cyzenis albicans* in 1955 (after Embree, 1966).

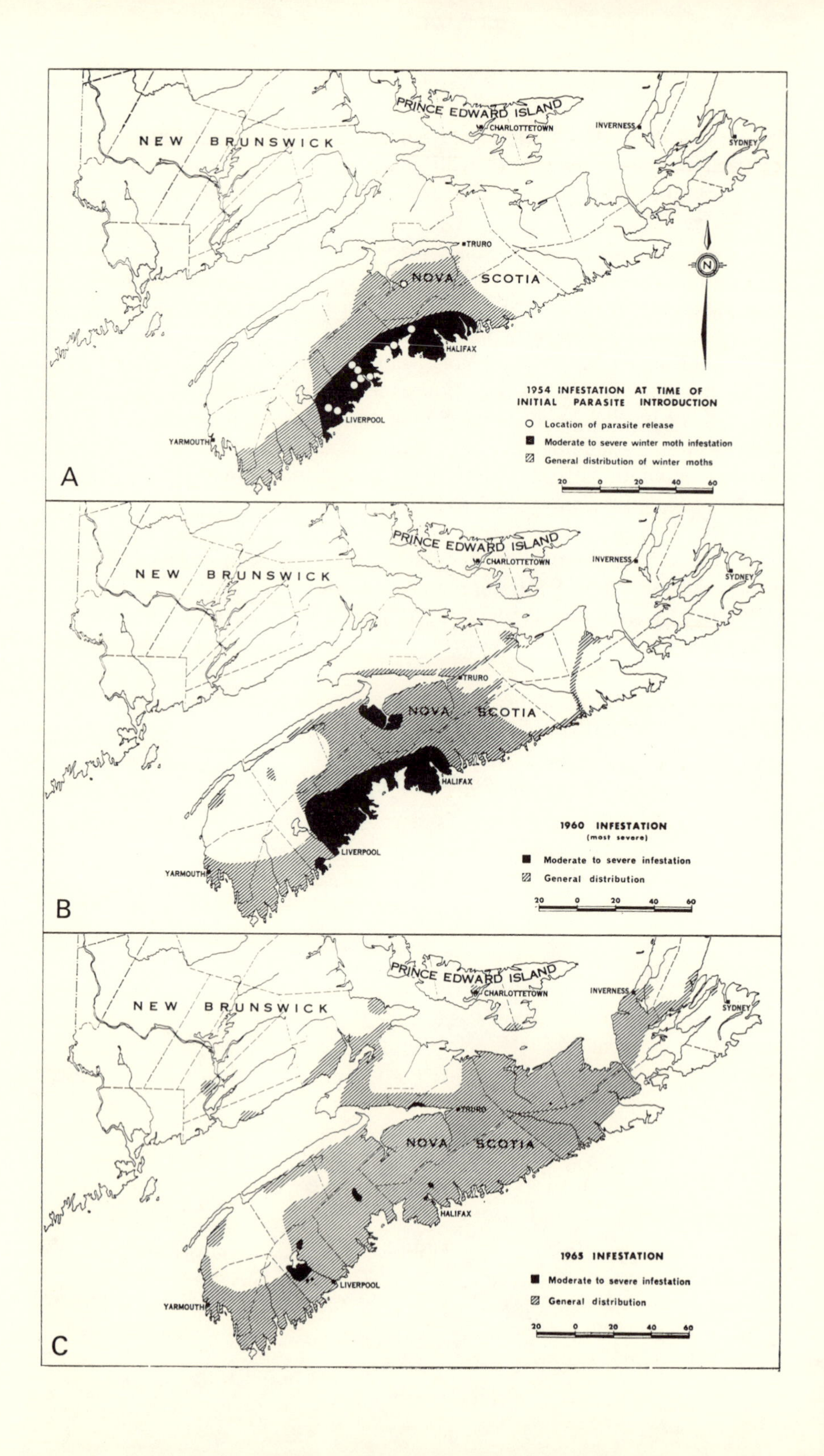

PRINCE EDWARD ISLAND
CHARLOTTETOWN
INVERNESS
SYDNEY
NEW BRUNSWICK
TRURO
NOVA SCOTIA
N
HALIFAX
LIVERPOOL
YARMOUTH
1954 INFESTATION AT TIME OF INITIAL PARASITE INTRODUCTION
Location of parasite release
Moderate to severe winter moth infestation
General distribution of winter moths
20 0 20 40 60
A
PRINCE EDWARD ISLAND
CHARLOTTETOWN
INVERNESS
SYDNEY
NEW BRUNSWICK
TRURO
NOVA SCOTIA
HALIFAX
LIVERPOOL
YARMOUTH
1960 INFESTATION
(most severe)
Moderate to severe infestation
General distribution
20 0 20 40 60
B
PRINCE EDWARD ISLAND
CHARLOTTETOWN
INVERNESS
SYDNEY
NEW BRUNSWICK
TRURO
NOVA SCOTIA
HALIFAX
LIVERPOOL
YARMOUTH
1965 INFESTATION
Moderate to severe infestation
General distribution
20 0 20 40 60
C

INTEGRATED PEST MANAGEMENT IN NORTH CAROLINA FLUE-CURED TOBACCO

The IPM program that has been developing in North Carolina flue-cured tobacco is a good example of how the management of insect, nematode, and plant pathogen pests can be effected jointly and in an environmentally and economically sound manner. By taking maximum advantage of available biological controls, cultural controls, and host plant resistance, and by utilizing economic thresholds in conjunction with a regular monitoring system, the North Carolina tobacco pest management program has minimized pesticide use while maintaining yields, cutting grower costs, and reducing environmental contamination.

North Carolina tobacco growers were plagued with many of the same problems that have led growers of other crops to seek an integrated approach to pest management. These included the rising cost of pesticides, pesticide residue levels in the harvested crop that were unacceptable to buyers, and the induction of the pesticide treadmill (Figure 8-6).

In this IPM program developed by R. L. Rabb, H. C. Ellis, F. A. Todd, and colleagues, the control of insect as well as several plant disease pests is largely centered on the prevention of pest population build up through the use of cultural controls and the enhancement of natural enemy activity. The cultural control program focuses on destroying food and refuges for overwintering pests. The strategy is to reduce overwintering and subsequent spring and summer pest populations to numbers that can then be kept below economically damaging levels by natural control factors.

The two major insect pests in North Carolina tobacco are the tobacco budworm (*Heliothis virescens*) and the tobacco hornworm (*Manduca sexta*). The budworm, by feeding on the developing leaf buds, inflicts damage early in the season before all leaves have emerged (buttoned) from the buds. The latter half of the growing season is devoted to expansion of already formed leaves, and the budworm, at any population density, has no significant economic impact. Therefore, after mid-July, when buttoning is completed, insecticide treatments for budworms are uncalled for and an unnecessary expense.

The tobacco hornworm, the other key insect pest, feeds on larger leaves and consequently may be a problem throughout the growing season, although it is generally more serious later in the season when leaves approach medium to full size. The last two instars of the hornworm are

FIGURE 8-5. Spread of the winter moth in Nova Scotia. The introduction of two parasites in 1955 is credited with slowing the rate of spread of this pest and with limiting the seriousness of its infestations (from Embree, 1966).

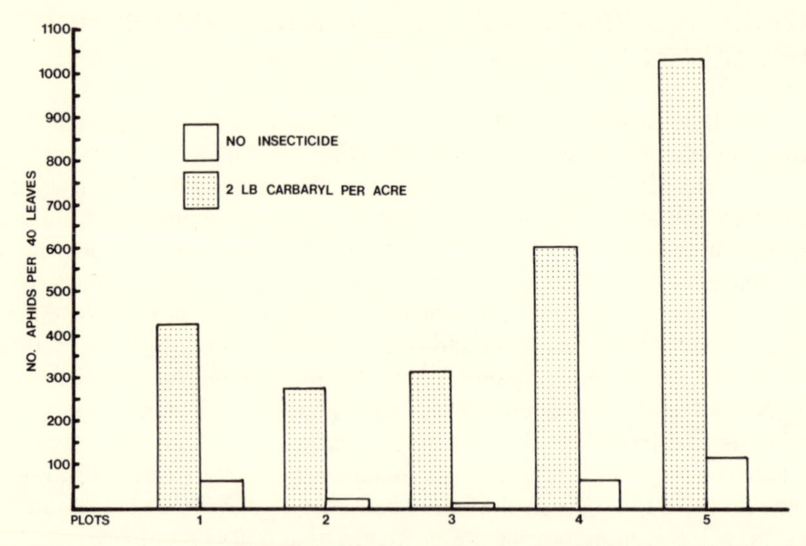

FIGURE 8-6. Secondary pest outbreak of the green peach aphid in North Carolina flue-cured tobacco. Note how aphid populations increase when an insecticide (carbaryl) is applied to control another tobacco pest and destroys natural enemies of the aphid (data from Rabb, 1971).

particularly damaging; a single voracious caterpillar may eat several large tobacco leaves during its last week of development.

In the past, insecticide treatments have often been made for flea beetle (*Epitrix hirtipennis*) control, but the economic status of this "pest" has not been clearly determined. Flea beetle populations may become quite dense early in the season soon after plants are transplanted from seeding beds into the field. However, since these first leaves are eventually dropped as the stalk develops, damage to them does not have a lasting effect. Flea beetle damage later in the season is purely cosmetic and does not significantly affect the grade or price of this kind of tobacco. But growers traditionally pride themselves on a crop of unblemished leaves, and it is difficult to convince them that such damage is not harmful. In the integrated pest management program, insecticide treatment for flea beetles is rarely, if ever, recommended. Aphids may become a problem in heavily sprayed tobacco fields. As Figure 8-6 illustrates, they are a secondarily induced pest, and natural enemies consistently keep aphids well below economic levels when pesticide use is minimized in the IPM program.

Nematodes, primarily root knot nematodes, are serious pests in many tobacco fields. If the grower has enough land, a rotation program can give effective control of these pests. Nematode-resistant varieties are available and these, in conjunction with cultural controls such as sanitation and rotation, are preferred over soil fumigation methods for nematode control in the IPM program.

Three plant pathogens are common tobacco pests in North Carolina. These include tobacco mosaic virus, bacterial wilt, and "black shank," a fungal disease. Resistant varieties are available for protection against all of these diseases, and cultural controls are also useful in their control.

A recognition of the important role naturally occurring biological control agents play in limiting insect pest populations has been the key to the success of the IPM program. To enhance the activities of these beneficial insects, insecticide treatments have been severely curtailed, and cultural controls are timed and carried out to complement natural enemy activity. Consequently, growers have been rewarded with an effective, free source of pest control. For instance, early season predation by the stilt bug (*Jalysus spinosus*), paper wasps, and a parasitic wasp (*Campoletis perdistinctus*) commonly negates the need for chemical control of the budworm in the northeastern part of North Carolina; in the southeastern part of the state, the budworm emerges a few weeks before stilt bugs are common in the fields, and here the pest may sometimes require an insecticide application. Likewise, natural enemies of the hornworm (including the stilt bug, paper wasps, a parasitic wasp (*Apanteles congregatus*), and a parasitic tachinid fly (*Winthemia manduca*) should keep the hornworm under sufficiently good control that not more than one (and often no) insecticide application is necessary during any season. Naturally occurring biological control is also responsible for keeping aphids and other potential insect pests well below their economic levels when unnecessary insecticide treatments are avoided in the integrated pest management program.

Cultural controls play an important role. As mentioned earlier, rotation and resistant varieties are key tools in the control of nematodes and plant pathogens. Postharvest sanitation, especially the removal of food sources and refuges for overwintering pests, is important in the control of insects as well as nematodes and plant pathogens. In addition, in-season removal of excess succulent growth can be instrumental in limiting late season or overwintering hornworm populations.

Tobacco is planted in special seeding beds in early spring and transplanted into the field as seedlings six weeks to two months later. This transplanting process can be critical in the spread of tobacco mosaic virus, which causes a serious tobacco disease. This virus may be spread mechanically on tools and with handling by the workers doing the transplanting. If tools and hands are periodically dipped in a milk solution, tobacco mosaic virus transmission can be kept to a minimum.

Since the tobacco plant is grown for its leaves, fruit and flower production is superfluous and an unnecessary drain of the plants' energy (Chapter 6, Figure 6-2) that could otherwise be directed to marketable leaves. For this reason growers have traditionally removed the flowering plant tops to keep the plants in the vegetative stage longer. However, topping stimulates another nutritional diversion from the tobacco leaves—sucker growth. So control of sucker growth has also become a standard cultural procedure.

It has been discovered that removal of the succulent tops and sucker growth is also beneficial in limiting hornworm populations. Succulent plants are attractive oviposition sites for adult moths and also provide the most desirable food for developing larvae. Late in the season, when most of the tobacco has been harvested, this lush sucker growth provides just about the only food source for hornworm larvae preparing to overwinter as pupae under the soil. If food is unavailable, the larvae starve and the overwintering population is greatly reduced, especially when a good postharvest sanitation program follows. Sucker control is carried out with contact chemicals applied to plant tops just when flower buds begin to form. A systemic sucker growth inhibitor is applied ten days later. This combination keeps suckers under control until stalk destruction takes place.

Postharvest stalk destruction is critical in limiting the number of hornworms and other pests that will be able to survive the winter and infest the fields the following spring. Stalks left standing until October or late September can provide the pests with enough food to complete the last larval stages. Then these mature larvae burrow deep into the soil to pupate and spend the winter in diapause (i.e., hibernation).

For this reason, in the tobacco IPM program, it is recommended that stalks be cut off and disked into the ground within a week after harvest. This procedure is then followed by a second disking two weeks later. A well-executed stalk destruction program is essential for good plant disease control as well. One undestroyed field may populate a whole county with hornworms the following spring; understandably, social pressure among neighboring farmers to get the stalks "out and under" is very strong.

Determination of the need for insecticide application for the hornworm and budworm is made with the assistance of experimentally derived control action thresholds in conjunction with monitoring data collected from individual tobacco fields. Each field is sampled weekly for hornworm, budworm, and natural enemy populations, as well as for two other insect pests, aphids and flea beetles. "Scouts" (as monitors are called in this program) also note the presence of disease symptoms, plant maturity, and dates on which various management practices have been carried out. This information is tabulated at a central university extension headquarters and sent out to growers along with suggested treatment levels. It is recognized that timing of insecticide application is critical. Insecticides applied too early or too late often do more harm than good.

Cost of the scouting program is about equal to the cost of one insecticide application. The IPM monitoring program can save growers at least this much in all areas of the state and much more in the Southeast where growers often spray up to seven or eight times a season, usually as "insurance" against pests. In certain tobacco growing areas, 75% of all insecticide applications could be eliminated with a carefully followed cultural and biological control program augmented by good scouting data.

Probably the greatest result of the pest management program has been educating growers to spray only when pests reach economically threatening levels rather than on a weekly or other programmed preventative basis. The idea has spread somewhat outside of North Carolina—yet most tobacco throughout the South is still sprayed according to rigid schedules. Despite this slow spread, where it is employed, the IPM program in North Carolina has met its goals by maintaining tobacco yields with sound pest control while considerably cutting costs and environmental pollution.

INTEGRATED PEST MANAGEMENT IN COTTON IN THE SAN JOAQUIN VALLEY OF CALIFORNIA

Cotton has been plagued with more pesticide-associated problems than perhaps any other crop. The positive side to this affliction is that the insecticide-inflicted problems of cotton have forced the development of a variety of integrated pest management systems in the crop around the world. Some of these have been referred to in earlier chapters, such as the Central American situation (Chapter 5) and the Texas situation (Chapter 7), where the use of a variety of cultural controls have markedly reduced the in-season use of insecticides for the control of the boll weevil and the pink bollworm in some areas, thereby permitting natural biological control of the major secondary induced pests—the cotton bollworm and the tobacco budworm.

The case history that follows involves the development of an integrated pest management program in another cotton-growing area of the U.S., the San Joaquin Valley of California, where more than 90% of California's cotton is grown. This system involves a different pest complex and, as might be expected, different sets of control action criteria and pest management tactics than systems in other cotton growing regions. The impetus to develop an integrated pest management approach in the San Joaquin Valley was precipitated by the usual "pesticide treadmill" crisis, i.e., pest resurgence (Figure 8-7), secondary pest outbreak (Figures

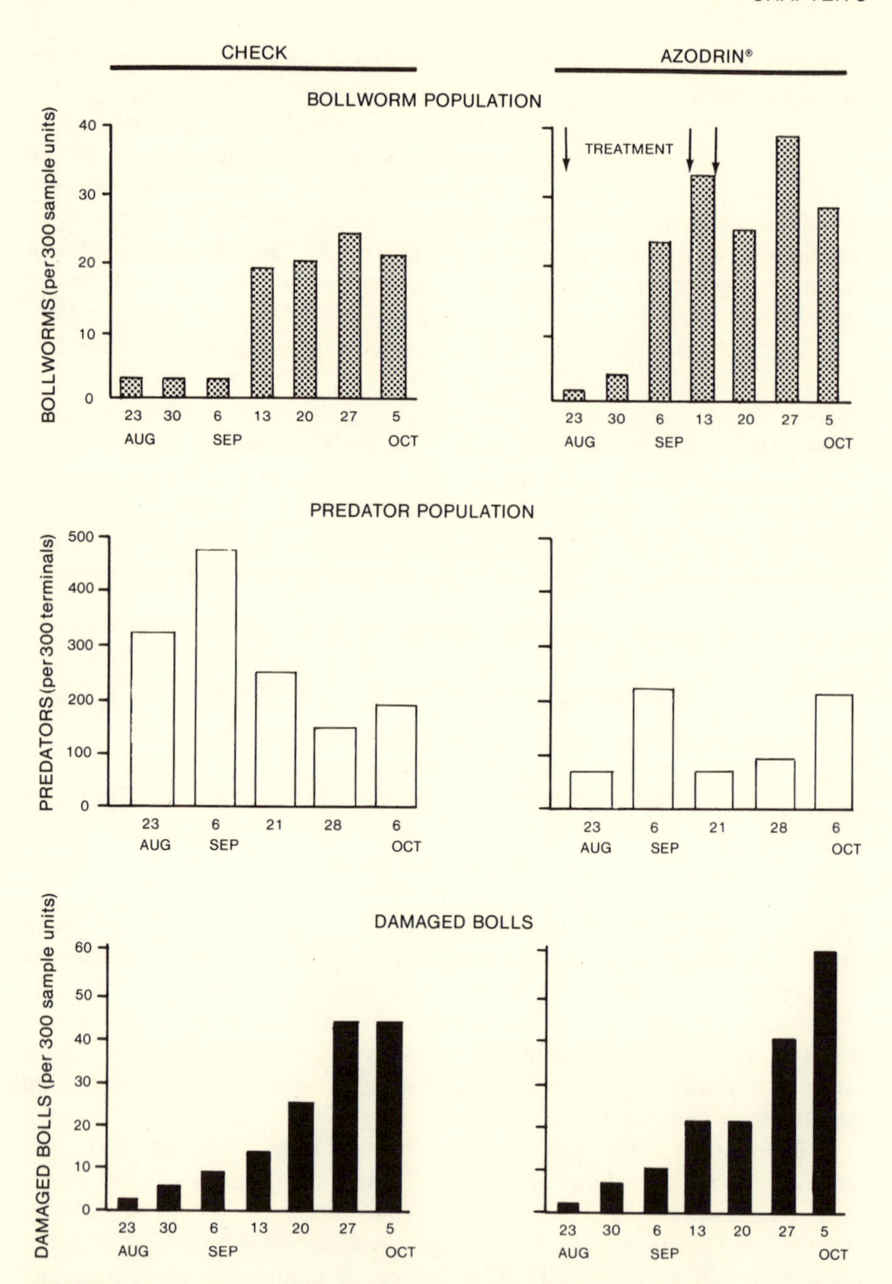

FIGURE 8-7. Resurgence of bollworms following treatment with Azodrin® in Dos Palos, California, 1965. Note especially the destruction of natural enemies in Azodrin® plots (vs. untreated check) and greater damage to bolls at the end of the season (from van den Bosch *et al.*, 1971).

8-8 and 8-9), and the development of insecticide resistant strains of major pests (Figure 8-10) followed by higher application rates, more kinds of pesticides, and more frequent applications.

The key pest in this cotton agro-ecosystem is the lygus bug (*Lygus hesperus*). Lygus is a plant bug that feeds primarily on the fruiting buds of the cotton plant, causing them to shrivel and drop off, thus prolonging the plant's vegetative state and delaying fruit (cotton boll) formation and ultimate yield. Historically, another major pest has been the cotton bollworm (*Heliothis zea*) that, as its name implies, attacks the bolls and also feeds on the plant terminals, buds, and blossoms.

The bollworm is probably the pest most feared by San Joaquin cotton growers because just a bit of feeding damage on a boll will allow entrance of rot-producing microorganisms that turn the boll into an unsightly, decaying mess. Current data indicate that this damage (unsightly as it may be) may not always be as serious as it looks; bollworms often feed on surplus flowers, buds, and small bolls that would not mature in any case. There is also abundant evidence that the bollworm in the San Joaquin Valley is essentially an insecticide-induced secondary outbreak pest that rarely rises to damaging levels unless its natural enemies are destroyed by chemical treatments.

Two other lepidopterous pests, the beet armyworm (*Spodoptera exigua*) and the cabbage looper (*Trichoplusia ni*), are also secondarily induced pests (see Figures 8-8 and 8-9), but they are not as widely feared as the bollworm.

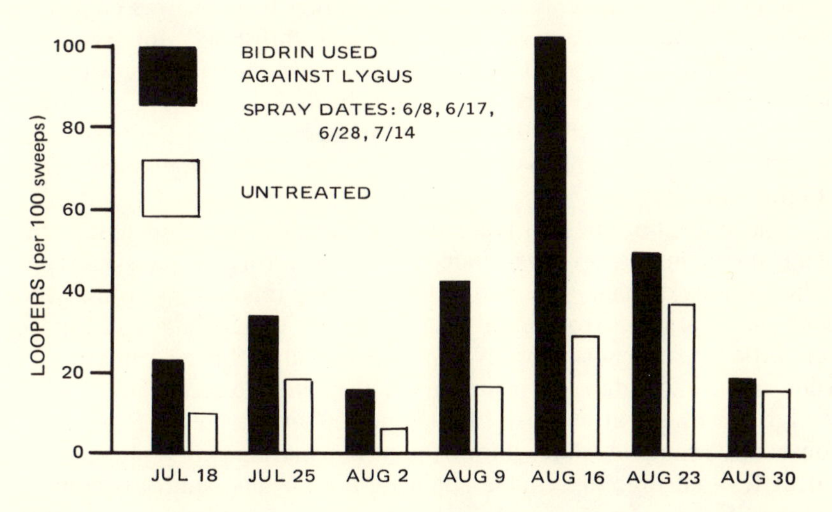

FIGURE 8-8. Secondary pest resurgence of cabbage looper following treatments of Bidrin® for lygus control, Five Points, California, 1966 (from van den Bosch *et al.*, 1971).

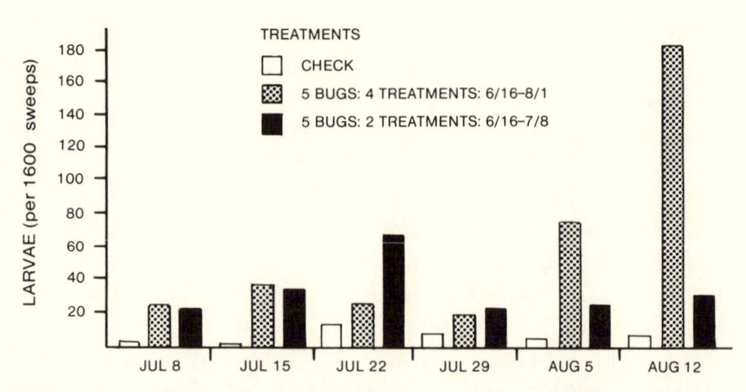

FIGURE 8-9. Secondary pest resurgence of beet armyworm in cotton following treatments of Toxaphene-DDT for Lygus control, Corcoran, California, 1969 (from van den Bosch *et al.*, 1971).

Since the bollworm problem appeared clearly to be the direct result of insecticide knockout of its important natural enemies, the integrated pest management program in California cotton sought to reduce chemical applications wherever possible. A second impetus toward this goal was the discovery that certain commonly used insecticides (e.g., methyl parathion, carbaryl) actually have a physiological effect on the cotton plant, resulting in decreased yields. The first step in the direction of reduced insecticide use was to understand precisely when the key pest, lygus, was economically damaging and at what population levels it was capable of producing economic damage. An "economic threshold" for lygus (and the bollworm as well) had been previously established, but with poor backup information. IPM researchers suspected that the lygus "threshold" was erroneous, since it remained static throughout the season and did not take into account changes in plant suceptibility to injury and degree of crop maturity.

Subsequent studies on lygus damage revealed that this assumption of a static economic level was indeed a serious error. In fact, lygus could inflict serious damage only during the squaring (budding) season, which in normal years runs from about the first of June to mid-July. Lygus poses very little threat to normally developing cotton after this peak of squaring. Thus, in the integrated pest management program, the economic threshold for lygus was set at at least 10 bugs per 50 sweep net sweeps on two consecutive sampling dates during the peak squaring period (June 1–July 15). After that time insecticide application for lygus control is avoided:

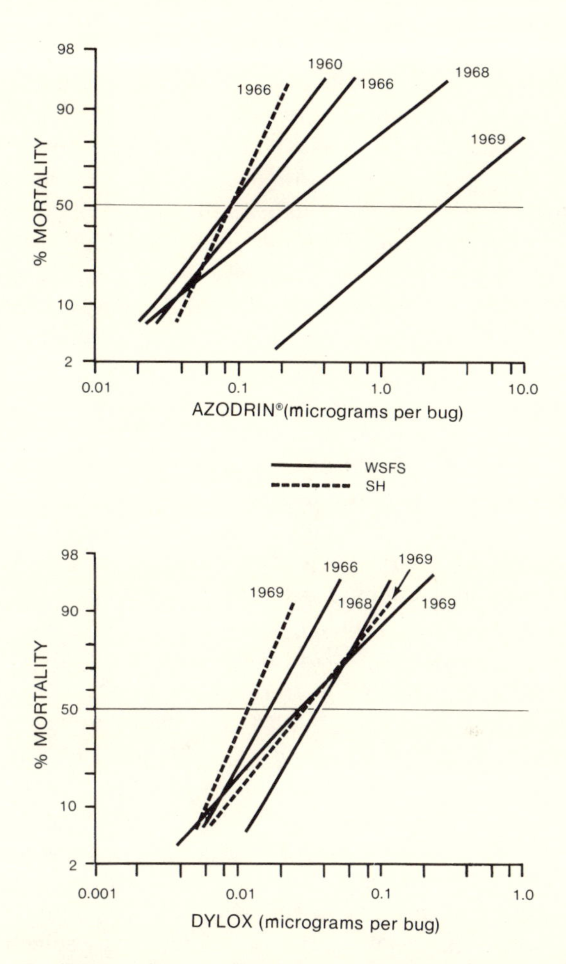

FIGURE 8-10. Resistance of *Lygus hesperus* to two organophosphorus insecticides in California. These data show that the bug is more resistant to the materials on the west side of the San Joaquin Valley (West Side Field Station, WSFS) than at Shafter (Sh) where less of these insecticides are used, and that in both places insecticide resistance is increasing (from van den Bosch *et al.*, 1971).

first because there is no reason for such applications, and thus it is a needless expense; and second because it can cause great damage to the crop by killing off natural enemies of the bollworm, beet armyworm, and cabbage looper just as their peak egg-laying period begins. Insecticide treatment at this time can (and frequently does) lead to serious outbreaks of all three of these lepidopterous pests. More recently in this ever-evolving IPM program an economic threshold that relates lygus populations to number of squares has been used. This measure, known as the lygus-to-square ratio, is more dynamic than the sweep net counts and more accurate, since lygus is less damaging as the cotton plant matures.

The key to a successful IPM program is a good field monitoring system. The sampling system in San Joaquin Valley cotton includes a reading of plant development as well as pest and natural enemy population counts. Plant records kept include plant height, percent defoliation, number of main stem nodes, and the fate of the fruiting points (for instance, have the buds, blossoms, and bolls dropped off because of insect feeding?; are the buds flowering?; are the bolls maturing?). Sweep net samples are used to survey lygus bugs, lepidopterous larvae, and predaceous insects. Four paired sets of samples are taken at four randomly chosen places in each field. At each site four 25-sweep subsamples are taken to assess pest populations, and four more 25-sweep subsamples are taken to gauge the abundance of natural enemies. Each field is sampled twice a week from the beginning of squaring (roughly mid-May) to the end of August. Additional sampling for cotton bollworm begins around August 1 and lasts until mid-September; this period is the peak of the bollworm damage season. At this time the upper one-third of all plants along a set row length are sampled for bollworms. The economic threshold for the cotton bollworm in untreated fields has been set at 20 small larvae (less than 1 half an inch long) per 100 plants. Thresholds for the other lepidopteran pests have not been set, but as a general rule of thumb, defoliation can be virtually ignored after August 1.

The most highly disruptive chemicals have been eliminated from the list of recommended materials; and in one case, a truly selective microbial insecticide, a nuclear polyhedrosis virus infectious only to the bollworm and the tobacco budworm, has been utilized in experimental programs. Use of such a highly selective material would be a very valuable addition to the IPM program as it is nontoxic to the important natural enemy component of the cotton field ecosystem.

In summary, the integrated pest management program in San Joaquin Valley cotton emphasizes a harmonious integration of naturally occurring biological controls and carefully timed insecticide treatments. The program is backed up by a well-designed monitoring system that records plant growth and maturity as well as pest and beneficial insect population levels and is commonly carried out by professional integrated pest management specialists (see Chapter 9).

INTEGRATED PEST MANAGEMENT OF MOSQUITOES IN A CALIFORNIA SALT MARSH

Mosquitoes are an excellent example of a group of pests whose successful control depends on an intimate knowledge of pest life cycle and biology. Mosquitoes are bothersome to people and domestic animals as adults, yet at the adult stage they are highly mobile and in many cases almost impossible to control. As larvae, mosquitoes are aquatic animals and are not pests; in fact, they serve a beneficial role as food for fish and other aquatic species. However, it is at the larval stage that mosquitoes are most vulnerable to environmentally sound modes of pest control. These methods focus on the elimination or modification of the standing bodies of water that may become mosquito breeding and development sites. "Source reduction," as the removal of mosquito breeding sites is known, may be carried out by draining, filling, pumping or temporarily removing water, or by removing the receptacles that may hold standing water (for example, old tin cans, old tires, or semipermanent puddles in unevenly graded irrigated pasture land). Mosquito control in permanent bodies of water such as ponds, irrigation ditches, sewage ponds, and marshes can be enhanced by encouragement and augmentation of natural enemy populations, especially mosquito-eating fish, and keeping the water in such situations circulating. The range of source reduction and biological control enhancement tactics for the control of mosquitoes in all their varied larval habitats is in most instances limited only by the innovative ability of the pest manager and the resources at hand.

Yet, despite the usefulness and dependability of the available "tried and true" source reduction techniques, for the last 30 years mosquito control has remained overwhelmingly dependent on chemical control. Insecticides for this purpose were at first cheap and effective, and their application was often easier than the laborious methods required to modify breeding sites. However, these chemicals are now failing to control mosquitoes in many situations because of the development of insecticide-resistant strains. In fact, in some areas disease-transmitting mosquitoes, and nuisance species as well, are resistant to a wide range of available larvacides. This problem, combined with a growing public awareness of the environmental hazards involved in the use of these materials, especially in aquatic ecosystems, has led pest managers to seek alternatives to chemical control. The following case history from Marin County, California, shows the utility of this approach.

The Petaluma salt marsh in Northern California forms an interface between the San Francisco Bay and Marin and Sonoma Counties' hills and pastures. For years the areas near the marsh were considered undesirable housing sites because they were populated by swarms of voracious mosquitoes generated in the wetland. However, suburban sprawl eventually approached the marsh area, and the new human arrivals demanded mosquito control. And they got action in the form of five aerial applications of the insecticide parathion yearly. Yet even with this massive dose of poison control was not satisfactory, and the Mosquito Abatement District received a continuous barrage of citizen complaints during the mosquito season.

The response to such complaints among the tax-supported California Mosquito Abatement Districts has traditionally been to apply another aerial spray of insecticide. But before instigating the integrated pest management program in Marin County, Dr. Alan Telford and his crew at the Marin Mosquito Abatement District (MAD) decided to investigate the biological aspects of the problem a little closer.

Since the San Francisco Bay's tide flushed through the marsh's maze of sloughs and channels twice daily, the wetland should have lacked the stability typical of a mosquito breeding area, and the high populations of these pests in the marsh simply did not make sense. However, a closer look revealed that the mosquitoes were actually breeding only in a special type of habitat provided by the marsh—these were "potholes" or "craters" that had been largely created by dummy bombs dropped in the area during World War II when the area was used as a practice bombing range. These holes, unconnected by drainage channels to the tidal sloughs, were not subject to regular flushing but instead existed as stagnant mosquito breeding sites that were recharged with water several times a year by unusually high tides. These potholes were ideal breeding sites for the salt marsh mosquito (*Aedes dorsalis*).

Mosquito breeding in larger permanent pools could usually be adequately controlled by natural enemies such as the mosquito-eating fish gambusia and an array of other important predators such as spiders, birds, and insects. However, populations of these natural enemies were repeatedly decimated by the frequent aerial insecticide applications, and mosquitoes were able to multiply in these pools unchecked by predators.

Telford and his group tackled the problem of eliminating the "bomb crater" breeding sites first. With the help of a special ditch-digging tool attached to an all-terrain vehicle, drainage ditches were established, connecting the craters to the main channels of the marsh so they could be subjected to the daily tidal flushing that kept the rest of the marsh mosquito-free.

Larger pools and ditches were stocked with gambusia fish, and these together with other mosquito predators (now able to survive in the absence of area-wide insecticide treatments) generally keep mosquito populations well below bothersome levels. Problems do occasionally arise, however, and mosquito populations in potential trouble spots are regularly scouted. Surveyors (or scouts) travel through the marsh in a specially adapted ditch buggy. This regular monitoring system, which samples for larvae in the breeding pools, allows prediction of mosquito flare-ups before the adults emerge. Thus, limited areas can be treated with a selective insecticide before the problem gets out of hand. The development of the ditch buggy has greatly facilitated the selective placement of insecticides to minimize environmental disruption. Today only two gallons of insecticide are used for mosquito control in the entire 2000-acre marsh.

Thus, the program has been a phenomenal success. Cost of insecticides alone has been reduced dramatically. The elimination of the costs of the aerial application procedure has provided an additional savings. The man-days required for mosquito control in the marsh have been reduced from 32 to 10 annually and will probably be further reduced. And most important of all, the marsh mosquito problem has disappeared from the nearby communities, and dairymen, whose cows graze around the marsh, have commented that their herds are free from tormenting mosquito swarms for the first time in memory. The marsh provides resting and breeding sites for many birds and other animals. The drastic reduction of pesticide use in the marsh has reduced hazards to these nontarget species, thus increasing environmental quality and enhancing the populations of many desirable animal species.

As a result, neighboring counties, eyeing the success of Marin County, have abandoned their pesticide-oriented control programs for salt marsh mosquitoes and are beginning to develop similar integrated mosquito control programs.

INTEGRATED PEST MANAGEMENT OF URBAN STREET TREE PESTS

Urban areas receive large quantities of pesticides; yet efforts to limit pesticide use in cities, where such toxic chemicals may pose serious health hazards to large human populations, have lagged far behind those in agriculture. Recently, several northern California cities have adopted integrated pest management programs for control of street tree pests and in doing so have drastically reduced pesticide use, saved taxpayers considerable expense, and maintained street tree quality. The following case history describes the program in the city of Berkeley where the first of these programs was initiated by William O. and Helga Olkowski.

Prior to the introduction of the integrated pest management program in the city of Berkeley, the manner in which pest control decisions were made was often haphazard. Although the city's street tree management personnel were well informed about tree and plant identities, they knew little about the pests on these trees and even less about natural enemies. Understandably, with such a small ecological and biological knowledge background, the city's street tree pest control crew rarely made an attempt to evaluate "pest" problems individually or to seek long-term biological, mechanical, or cultural control solutions. Instead, they "necessarily" relied on what seemed to be a quick, easy method requiring no biological input—pesticide chemical application at the first suggestion of trouble. For instance, when a citizen called up to complain about a "sick" tree in front of his/her house, a city crew would be sent over to spray the tree or a block of trees in the area, often with little or no examination of pest populations. As a result, large quantities of pesticides were applied unnecessarily.

Thus, the first step in implementing the IPM program was to change the city personnel's response to a citizen complaint from one of automatically applying a pesticide to the designated tree or tree area to a policy of carrying out a thorough inspection of the tree to determine if a control action is really needed.

Pest damage to street trees is rarely severe enough to produce economic damage (that is, destroy the tree), yet pest populations at certain densities cause damage that is unsightly or otherwise bothersome to large numbers of residents. Therefore, the pest control action threshold (or, as it is often called, the aesthetic injury threshold) in such an urban tree situation is the pest population level at which the annoyance (or losses) caused by the pest exceed the cost and/or undesirable side effects of applying control measures. Such aesthetic injury levels and action thresholds are even harder to establish than economic thresholds because different people have varying tolerances for pest populations on street trees. After learning that such damage does no permanent injury to the tree, citizens will often tolerate a much higher level of aesthetic injury for short periods of time. Consequently, educating the public to these facts is a large part of the urban street tree pest manager's job. When disagreements arise as to what constitutes "aesthetic" injury in the IPM program, it is always best to make control action decisions which favor minimized pesticide use.

Aphids caused the most common insect problems in Berkeley's street trees. Large populations of aphids, lacking effective parasites or freed from control by natural enemies because of insecticide overuse, would develop on trees in the spring and summer. These aphids, which are plant sap feeders, would excrete heavy amounts of a sticky liquid substance known as "honeydew." The honeydew dropped on cars, vegetation, sidewalks, houses, and people causing an unsightly sticky mess, sometimes attracting flies or stinging wasps, and providing an environment for growth of an ugly black "sooty mold" or fungus. Honeydew-engendered concerns made up the largest category of complaint calls to the Berkeley City Parks and Recreation Department.

Naturally occurring biological control agents for these aphid pests were locally resident in Berkeley and could be very effective in many situations if not killed off by heavy insecticide use. These resident natural enemies were mostly predators and included lacewings, lady beetles, and syrphid fly larvae. However, the frequent use of toxic pesticides over a period of years had decimated their populations. In addition to already existing biological control agents, parasites of some of the most serious aphid pests (i.e., the linden aphid, the elm aphid, and two aphids on English oak) were imported from Europe by the University of California's Division of Biological Control and released in the city of Berkeley.

In order to protect and encourage both the naturally occurring and introduced natural enemies, insecticide applications in the street tree ecosystem had to be minimized. This was done in a number of ways.

First, a microbial insecticide, *Bacillus thuringiensis* ("B.t."), replaced "conventional" chemical insecticides for control of another key pest, the oak moth caterpillar, *Phryganidia californica*. B.t. infects and kills only certain caterpillars and is harmless to most natural enemies, man, and other nontarget species. It is sprayed according to the same kinds of control action criteria and with the same kind of equipment used for chemical insecticide application yet avoids outbreaks of secondary pests by preserving natural enemy populations.

Various cultural, mechanical and physical controls were useful in augmenting the action of natural enemies in the control of the tree-infesting aphid pests. Aphid populations frequently concentrate in the protected inner canopy of the tree and cannot survive as well on the more exposed outer edges where they are more severely subjected to natural control factors such as weather (e.g., wind, rain, cold, heat) and perhaps also natural enemies. Thus by pruning out this "aphid haven," especially when populations are high, aphid problems can be reduced considerably.

Another aphid control tactic is to keep ants out of the trees. Ants, the notorious aphid diarymen, protect these pests from natural enemies and distribute them around the tree for optimum production of the aphids' sweet honeydew secretions. If there is a trail of ants going up into a "problem" tree this is a clue to the cause of the problem. Application of a sticky ring (see Chapter 4, Figure 4-8) around the trunk will prevent ants from getting up into the foliage from their underground nests and will thereby lower aphid populations and unsightly honeydew drip and the sooty mold it leads to.

Like other IPM programs, the urban street tree program depends on careful monitoring to predict and evaluate pest population build-up and biological control potential, and to justify or select control measures. In the Berkeley program, trees in known "trouble" areas are regularly monitored for pest build up as are problem trees brought to the city's attention by concerned citizens. When monitoring indicates that pest populations have reached aesthetic threshold levels and that naturally occurring biological control factors will not be able to restrain or lower the population, a control action is taken. This may be the application of B.t. for lepidopterous pests such as the oak moth, or water sprays or water and soap solution sprays for aphids, leafhoppers, and plant bugs. Such sprays knock pest species off the tree and usually keep their numbers down until late spring or early summer when predators and parasites are able to keep these pests in check. As Table 8-2 indicates, no other pesticides have been required in the Berkeley IPM program in recent years.

An important component of this pest management program is education; this includes the education of both city workers and city residents. The street tree crews, but most particularly persons in decision-making positions, must be able to recognize beneficial as well as pest insects and to distinguish insect damage from plant disease symptoms and "sick" trees suffering from overwatering, underwatering or nutritional deficiencies, and other environmental stresses. Good monitoring techniques are essential and decisions must be made with an understanding of aesthetic thresholds and the effects of various control actions on nontarget species, including that common urban inhabitant—man. Above all, the city pest management employees must not be too eager to treat and must be able to communicate the value of the integrated pest management philosophy to the people they serve, the city's residents.

Citizen education is critical to urban pest management. Many urban and suburban residents are afflicted with a fear of insects, a phenomenon known as *entomophobia,* which makes them intolerant of any population level of insects—pest *or* beneficial. Others are ignorant of the role of beneficial insects and assume all insects are pests. Most cannot distinguish pest damage from other symptoms and often genuinely fear that a tree is dying if it begins to drop leaves or rain down honeydew. All need to be given mini-courses about the urban environment they live in; they need

TABLE 8-2
Insecticides Used by the Department of Recreation and Parks, Berkeley, California, 1969–1975[a]

Insecticides	Toxicity (LD_{50})[b]	Amounts used[c]						
		1969	1970[d]	1971[e]	1972[f]	1973[g]	1974	1975
DDT	150 mg/kg	65	0	0	0	0	0	0
Diazinon	250 mg/kg	16	6	2.5	1	48 oz	0	0
Dimethoate (Cygon)	215 mg/kg	9	7	3.7	2	0	0	0
Malathion	1156 mg/kg	12	7	1.9	0	0	0	0
Meta-Systox R(Thiometan)	40 mg/kg	2	0	0	0	0	0	0
Kelthane	800 mg/kg	2	0	0	0	0	0	0
Aramite	2000 mg/kg	0	3	0	0	0	0	0
Thuricide	Harmless	0	30	3.0	21	2	0.5	7.5
Chlordane	283 mg/kg	0.5	0.5	0.5	0	0	0	0
Lead arsenate	100 mg/kg	200 lb	0	0	0	0	0	0
Carbaryl	500 mg/kg	60 lb	20 lb	0	0	0	0	0
Lindane (gamma isomer)	125 mg/kg	0	1	0	0	0	0	0
KXL[h]	High	0	0	3.3 qt	0	0	0	0

[a] From Olkowski *et al.* (1976).
[b] Abbreviation for median lethal dose; indicates the amount of substance that kills 50% of animals tested (usually rats, orally administered) expressed in weight of chemical per unit weight of animal (mg/kg). The higher the number, the less toxic is the insecticide.
[c] Submitted to authors by Department of Recreation and Parks. Percentages of active ingredients unknown. All numbers are in gallons except those noted.
[d] Note the drop in pesticide use following the instigation of the program in 1970.
[e] Data taken from monthly summaries of pesticide use report; herbicides and 23.6 gal of Zineb also used.
[f] 24 gal Zineb and herbicides also used.
[g] Ounce abbreviation means fluid ounces (128 fl oz = 1 gal).
[h] A trade name for a mixture that contains vegetable oil, copper oleate, pyrethrum, and rotenone; this material is not used on shade trees in Berkeley.

to learn about the roles of natural enemies and pests and about the trees' natural defenses against such pests. They also need more information on the negative side effects of the use of toxic substances in any urban environment. Usually a little ecological education will greatly increase citizens' tolerance of insects and convince them that the minimum use of hazardous materials is a desirable goal in their city. Such citizen education can occur at all levels of public school instruction, in local newspapers, garden club talks, community meetings, via handouts given to curious residents by the street tree pest managers as they carry out their work, and through a host of other communication channels.

The Berkeley program, which manages over 30,000 street trees (123 species) over about 110 acres, has provided an excellent example of the benefits integrated pest management can bestow on urban and suburban

TABLE 8-3
Pest Populations of Experimental Apple Orchard Plots, 1950–1962

Pest species	Pest status	Type of injury	Biological control	Chemical treatment	Population monitoring	Pest ranking
California red scale, *Aonidiella aurantii* (Maskell)	Major	Direct and indirect—Infests fruit, twig dieback, defoliates	Ineffective	24-Month interval; 30- to 36-month interval (occasionally)	Sex pheromone traps	1
Citrus thrips, *Scirtothrips citro* (Moulton)	Major	Direct—Cosmetic	Ineffective	Petal fall	Fruit counts	2
Cutworms, *Xylomyges curialis* Grote	Major	Direct—Young fruit destroyed or badly scarred	Ineffective	March–June, except during bloom period	Foliage shaking Light traps	3
Citricola scale, *Coccus pseudomagnoliarum* (Kuwana)	Major	Direct—Leaf and twig dieback Indirect—Cosmetic, sooty mold	Ineffective	24- to 36-Month interval	Leaf counts	4
Fruit tree leaf roller, *Archips argyrospilus*	Major	Direct—Leaves and young fruit	Ineffective	April, except during bloom priod	Leaf inspection	5

furcata Brunner von Wattenwyl		destroyed or badly scarred				
Citrus red mite, *Panonychus citri* (McGregor)	Minor	Indirect—Fruit Direct—Leaf stripping	Effective some years: predaceous mite, virus disease	Rarely treated	Leaf counts from permanent sample trees	7
Yellow scale, *Aonidiella citrina* (Coquillett)	Minor	Direct—Leaves and fruit Indirect—Cosmetic on fruit	Effective	Rarely treated	Leaf counts (laboratory)	8
Cottony cushoin scale, *Icerya purchasi* Maskell	Minor	Indirect—Cosmetic, sooty mold, leaves and fruit	Effective	Rarely treated	Tree inspection	9
Aphids, various species	Minor	Direct—Leaf curling Indirect—Sooty mold, cosmetic	Effective in conjunction with high temperature	Rarely treated	Tree inspection	10
Brown soft scale, *Coccus hesperidum* Linnaeus	Minor	Direct—Leaf and twig feeding Indirect—Sooty mold, cosmetic	Effective	Rarely treated	Tree inspection and leaf counts	11
Leafhoppers	Minor	Direct—Fruit feeding	Unknown	October, November	Tree inspection	12

TABLE 8-4
Pesticide Usage Priorities for Control of Citrus Pests in the San Joaquin Valley

Pest	Pesticide	Preference
California red scale[a]	Carbaryl and oil	1
	Parathion	2
	Supracide (new)	3
Citrus thrips	Ryania	1
	Parathion	2
	Dioxathion	3
	Dimethoate	4
Cutworms	Trichlorfon	1
	Parathion	2
	Carbaryl	3
	Naled	4
	Lannate	5
Fruit tree leaf roller	Trichlorfon	
	Bacillus thuringiensis	
Katydid	Parathion	
Citrus red mite	Morestan	

[a] Pesticides used for California red scale also control citricola scale.

fornia growers, cannot be used in the San Joaquin Valley because of greater phytotoxicity brought on by climatic factors, some preferences can be given certain materials because of their less intensive disruption of the activities of natural enemies in the citrus grove. These are shown in Table 8-4. Note, however, that these preference ratings have not taken into consideration human, bee, or other nontarget organism toxicity.

INTEGRATED PEST MANAGEMENT IN PEARS

An integrated pest management program for the control of insect and mite pests on pears was funded by the USDA APHIS program and developed by the University of California Cooperative Extension under the leadership of Dr. Clancy Davis during the four-year period 1973–1976. At the same time, monitoring programs for two plant pathogen pests and a model for predicting codling moth outbreaks were being developed independently by University researchers. All have been of benefit to growers and have been incorporated into an integrated pest management program now delivered by independent pest control advisors to growers in various areas of northern California.

Of all the deciduous fruits, pears probably presented the most difficult pest control problems for orchard growers in California. Codling moth was a perennial pest, and multiple treatments applied for its control inevitably resulted in severe destruction of natural enemy populations and subsequent flare-up of pest aphids, scales, psylla, and phytophagous mites. Mites—largely two spotted mites, European red mites and rust mites—were a particularly serious secondary outbreak pest problem, as pears are extremely sensitive to mite damage.

Prior to the instigation of federal funding, research had indicated that an integrated pest management program had promise. In fact, a minority of growers and private pest management consultants had already begun to realize the benefits of a healthy natural enemy complex in their orchards with the introduction of new techniques involving selective dosages and better timing of insecticide applications. Earlier research had shown that one-half normal rates of azinphosmethyl (Guthion®) could give effective control of the codling moth and allow the survival of natural enemy populations.

Critical to the use of this low dosage rate has been careful timing of applications. For the most efficient use of codling moth treatments, materials must be applied when moth larvae are in the tiny first instar stage, before they bore into the fruit, where they are protected. Applications that are timed according to phenology of the tree or time of the year (as had often been the case in the past) are apt to miss this very short but critical period and result in poor kill—leading to the emergence of substantial numbers of moths in the next generation and thus making further spraying necessary to keep the pest below economically damaging levels throughout the season.

The earliest program utilized virgin-female-baited traps (Figure 7-3C) to determine time of peak adult flights. Since codling moths occur in fairly well-synchronized broods, treatable populations of the vulnerable newly hatched first instar larvae could be expected 2–4 weeks later (depending on weather). Later, virgin female traps were replaced with synthetic-pheromone-baited traps. Frequent visual observations of the developing eggs in the field can give the pest manager an even more precise idea of when to treat.

Timing of codling moth sprays has been further facilitated by the development of a computer-based model by L. A. Falcon, Carolyn Pickel, and associates that incorporates weather and pest sampling data to follow moth phenology and determine optimum treatment times. The most sophisticated pest managers maintain weather stations equipped with hygrothermographs to get weather data from individual orchards. Growers in the program have been successful in reducing their codling moth applications by using these monitoring tools.

The subsequent decrease in pesticide use, especially of materials that are more disruptive to the natural enemy complex than the low rates of azinphosmethyl, has greatly decreased mite problems. In fact, in the Sacramento Valley, area mite treatments (with the exception of oil applications) are now often not needed. This reduction in acaricide use will undoubtably slow the development of resistance among this rapidly adapting group of arthropods. Pears in other areas, such as Lake County, have more severe mite problems due mainly to climatic factors.

Two other insects have the capability of becoming major problems if proper management programs are not instigated. These are the San Jose scale and the pear psylla. Both can be controlled by the application of dormant oils during winter months in most parts of California. These treatments are an integral part of the integrated pest management program. If not properly executed, in-season treatments will be required to control these pests. Such summertime treatments are extremely disruptive to mite predator activities and to parasites of scales and other pests. Inadequate dormant oil coverage can result also in outbreaks outside the negligent grower's orchard due to orchard-to-orchard migration. This has been a major problem with the pear psylla. Growers in one county became so concerned about this reoccurring situation that local authorities have instituted a regulation requiring dormant applications in all infested orchards.

Good weed management is important in manipulating favorable predator to prey mite ratios; weeds such as perennial morning glory climbing up tree trunks can produce an environment favorable to phytophagous mites. On the other hand, a healthy cover crop between rows of trees may enhance predator and parasite activity. Consequently, herbicide applications are made directly around the base of the pear trees while weeds in interrow areas are allowed to grow and are periodically mowed.

The plant pathogen monitoring aspect of the pear pest management program has been particularly innovative. The most serious disease plaguing pear growers is fire blight, a bacterial disease that enters the tree through the blossoms. Previously, growers automatically sprayed orchards every five days or so during bloom with a pesticide such as Bordeaux mixture or streptomycin to control possible infections of this disease. If growers waited until fire blight symptoms were apparent in their trees to determine if treatments were necessary, it would be too late to save many trees from this devastating disease with pesticide chemicals. At this time, drastic pruning would be necessary, a costly and time-consuming operation, and even with this, crop loss would occur.

However, a new sampling technique has permitted elimination of many of these "preventative" applications. Developed by Miller and Schroth at the University of California, the technique utilizes a culture medium selective for the growth of the fire blight bacterium. Nectar samples taken from orchard blossoms are dropped onto the culture media; presence of fire blight is detected by growth of colonies on the culture plate. The test has been made even more useful by the incorporation of streptomycin in some cultures to detect streptomycin-resistant strains of the bacteria in the field. Similar monitoring programs have been used for scab, another disease problem in some apple growing areas.

The integrated pest management program in pears has been a great success. One gauge of this success is the number of growers who were willing to pay the cost of employing private consultants to continue the program after the federal funding had ended. Approximately 50% of the total pear acreage in the initial USDA-funded IPM program continued their participation with private consultants, and now about 20% of the total pear acreage in California is under integrated pest management. In general, growers were able to more than pay for monitoring costs through savings in insecticide costs, although savings varied from area to area depending on severity of pest problems.

The pear pest management program is also an ideal model of how new integrated pest management programs might be initiated in other crops. Growers at first paid only half of the monitoring fee. In subsequent years growers paid a larger and larger portion of these costs as they began to recognize the program's benefits. In addition, the program gave basic training and provided grower contacts for several pest managers who became independent pest control consultants in the area as federal funding was phased out. Getting established, gaining grower confidence, and getting good field experience are real stumbling blocks for pest managers who are trying to set up a consulting business; in this case the groundwork was done for them by a trusted institution, the University Cooperative Extension.

CHAPTER 9

The Integrated Pest Management Specialist

One reason conventional chemical control methods (i.e., pesticide applications according to a predetermined calendar schedule or without use of monitoring and regular control action criteria) have been so popular and are often only reluctantly abandoned by growers, government personnel, homeowners, and others having pest problems is the relative ease with which they can be used. Just about all that is required for their use is acquaintance with the application equipment and safety precautions and the ability to read a label or a calendar.

Integrated pest management requires professional attention. Those making pest management decisions according to IPM dictates must have a solid biological and ecological background embracing both efficacy of control tactics and their indirect or direct on-site and off-site consequences. They must have a knowledge of the pests, natural enemies, and crop or resource phenology. They must understand how the abundance and life cycle of each key element in the resource system changes throughout the year according to changes in weather, natural enemy populations, or management practices. Integrated pest management decisions are made according to data collected in regular, systematic field monitoring programs and as determined by established control action criteria. Thus they require the establishment of a regular field monitoring program and decision-makers well-acquiainted with economic, aesthetic, or other control action criteria. The techniques, methods, strategies, and control action criteria used in IPM programs are constantly being improved, thus decision-makers must keep abreast of these changes and be constantly upgrading their own field programs.

However, the responsibilities facing today's growers, foresters, city parks personnel, or other resource managers are many. Controlling pests is only a minor part of their task, and they usually have little time to keep up with new developments in pest management. Thus, the temptation to apply chemicals at the first sign of pest activity or according to a predetermined application schedule as "insurance" against damage is great. Moreover, the immediate utility of a chemical to lessen a perceived potential problem is, of course, appealing.

A recent USDA report noted that farmers were not using the integrated pest management approach in pest control decision-making because it required basic changes in five attitudinal or human behavioral patterns: (1) growers were not accustomed to using control action thresholds as decision-making criteria to time pesticide application but instead traditionally followed "preventative" predetermined application schedules; (2) growers were not accustomed to using regular pest *monitoring* programs—in the past, field checks were random and irregular, if made at all; (3) growers were not accustomed to making pest control decisions in a *dynamic* context (i.e., considering a number of ecosystem variables—for instance, crop stage, natural enemy populations, weather)—they always relied on routine spray schedules with no consideration of these ecosystem variables; (4) growers were not accustomed to using control measures *other than chemical pesticides*; and (5) growers were not accustomed to the idea of controlling pests on an *area-wide basis* (e.g., the cotton stalk destruction programs for boll weevil and pink bollworm in Texas cotton) but had always considered pest control as the individual farmer's problem.

The report further pointed out that in most communities, farmers have neither the technical knowledge nor the time to implement integrated pest management practices by themselves and would require the services of professional integrated pest management specialists to carry out these programs. But this is an entirely new concept to most growers, and they are not as yet willing or able to grasp its significance, especially as regards their welfare. They tend to view it as something they should do for someone else's benefit. On the other hand, such integrated pest management services are being provided and accepted by growers across the country for a variety of agricultural crops and for urban pest management situations. Resource managers or growers enlisting such services get systematic, timely field monitoring and management counseling on when control action is needed and what alternatives are available. In addition, they often receive a great deal of other advice on production management problems ranging from choosing plant varieties to the strategic timing of planting, irrigating, fertilizing, harvesting, and cultivation procedures for optimal resource return. Many pest management specialists make it their top priority to stay informed about the latest methods and techniques and thus keep their clients' crops under the most up-to-date protection program available. The grower or resource manager can then concentrate his time on other responsibilities involved in the production of the resource—things closer to his own competence and experience—confident that the crop/resource is getting the best pest management system available. Of course, if not satisfied, he can take his business to a competing pest management specialist the following year.

What is the cost of using such integrated pest management specialists? IPM programs, employing IPM specialists, generally save growers as much or more money in insecticide application cost (as opposed to conventional chemical control) than the cost of their services. For instance, a detailed study by Hall, Norgaard and True (Tables 9-1, 9-2, 9-3, and 9-4) showed that cotton and citrus growers in the San Joaquin Valley of California who used the services of IPM consultants saved money on insect control costs (including consultant fees) in both crops.

Integrated pest management specialists may be employed by state extension services as part of government-supported or area-wide IPM programs, employed by grower-owned organizations, self-employed, or work with a group of specialists offering private consultant pest advisory services to individual growers.

Although government-financed extension IPM programs have been instrumental in showing the feasibility of IPM in a range of crops, with the limited funding available through public agencies, they cannot possibly reach the majority of growers in need of such services. Their advantage is that they have provided free or very inexpensive services to somewhat wary growers in various IPM pilot programs. Since growers are apprehensive at first about trying out new systems, the low cost factor has been important in enlisting their initial cooperation and acceptance. (However, once the benefits of the pilot programs have been revealed, it is important that individuals in the private sector be encouraged to offer these services so that extension personnel can continue to demonstrate new programs that arise from the research efforts.)

Private IPM specialists may be hired by grower cooperatives, individual growers, municipalities, homeowners, or any individual or group requiring regular pest management services. Sometimes, large growers or grower groups may hire an individual for full-time employment as their IPM specialist. Most often, however, the integrated pest management specialist contracts with a number of different growers each season. Services may range from simple insect monitoring and insect control advisory services offered by individual specialists running a one-person/one pickup truck operation, to complete crop management services offered by crop consultant groups whose personnel include individuals trained in a wide range of crop production and pest management specialties.

Groups that offer complete crop management services are likely to

TABLE 9-1
Average Dollar Yield Per Acre[a] in San Joaquin Valley Cotton, 1970–1971[b,c,d]

Cotton acres	Nonusers 1970	Users 1970	Nonusers 1971	Users 1971	Nonusers 1970–1971	Users 1970–1971
0–49	247.50	228.75	264.60	292.95	254.80	259.00
	(70.87)	(28.01)	(98.32)	(28.19)	(82.02)	(40.94)
50–199	241.25	216.25	311.85	250.43	273.00	236.60
	(34.18)	(33.72)	(52.53)	(20.80)	(54.26)	(35.12)
200–999	272.50	251.25	321.30	281.93	295.40	264.60
	(39.43)	(36.42)	(42.58)	(40.21)	(47.18)	(40.42)
1000 or more	251.25	285.00	192.15	281.93	232.40	274.40
	(0.78)	(23.76)	(50.40)	(9.17)	(46.53)	(16.89)
Average[e]	255.00	271.25	221.65	281.93	247.80	270.20
	(22.71)	(35.38)	(72.93)	(20.51)	(55.08)	(27.49)

[a] These figures are based on a 1970 price of $0.25 per pound and a 1971 price of $0.31 per pound. The 1970–1971 price was $0.28, an average of the two years.
[b] From Hall *et al.* (1975). Data collected from Willey-Norgaard research for the Ford Foundation.
[c] Standard deviations are indicated in parentheses.
[d] "Users" are growers using independent pest control consultants' services; "nonusers" are growers not using consultant services.
[e] These statistics are not calculated by averaging the numbers given in this table. They are calculated using total acres and total yield for each category.

TABLE 9-2
Average Insecticide Cost Per Acre in San Joaquin Valley Cotton, 1970–1971[a,b,c]

Cotton acres	Nonusers 1970	Users 1970	Nonusers 1971	Users 1971	Nonusers 1970–1971	Users 1970–1971
0–49	22.01	6.53	23.24	13.02	22.65	10.02
	(6.92)	(5.52)	(10.67)	(9.02)	(9.09)	(8.26)
50–199	13.17	6.49	14.83	9.02	13.93	7.34
	(10.26)	(3.49)	(8.19)	(3.21)	(9.41)	(3.67)
200–999	14.10	9.13	13.62	8.92	13.87	9.02
	(7.77)	(5.07)	(9.95)	(7.08)	(8.90)	(6.23)
1000 or more	7.84	4.65	15.54	2.79	11.25	3.41
	(2.92)	(2.05)	(7.25)	(2.33)	(6.79)	(2.38)
Average[d]	9.34	6.13	15.16	4.21	11.97	4.94
	(5.51)	(4.61)	(8.06)	(4.72)	(7.38)	(3.85)

[a] From Hall *et al.* (1975). Data collected from Willey-Norgaard research for the Ford Foundation.
[b] Standard deviations are indicated in parentheses.
[c] "Users" are growers using consultant services; "nonusers" are growers not using consultant services.
[d] These statistics are not calculated by averaging the numbers given in this table. They are calculated using total acres and total insecticide costs for each category.

TABLE 9-3
Average Dollar Yield Per Acre in San Joaquin Valley Citrus Orchards, 1970–1971[a,b,c]

Citrus acres	Nonusers 1970	Users 1970	Nonusers 1971	Users 1971	Nonusers 1970–1971	Users 1970–1971
0–25	252.79	477.50	396.10	505.64	324.45	491.57
	(55.81)	(72.08)	(142.83)	(251.39)	(129.97)	(185.45)
26–100	449.72	390.33	472.44	517.83	461.08	453.32
	(117.38)	(167.14)	(136.04)	(227.20)	(161.69)	(209.04)
Over 100	545.15	561.19	510.37	504.72	527.76	529.12
	(180.22)	(243.35)	(104.03)	(286.60)	(125.15)	(269.72)
Average[d]	509.47	527.17	496.23	506.65	502.85	515.80
	(187.36)	(237.22)	(118.79)	(274.81)	(157.00)	(260.64)

[a] From Hall *et al.* (1975). Data collected from Willey-Norgaard research for the Ford Foundation.
[b] Standard deviations are indicated in parentheses.
[c] "Users" are growers using consultant services; "nonusers" are growers not using consultant services.
[d] These statistics are not calculated by averaging the numbers given in this table. They are calculated using total acres and total yield for each category.

give the most all-inclusive pest management advice, for theirs tends to be a more holistic service and therefore more ecologically relevant. These groups make recommendations for fertilizing, irrigating, selecting desirable plant varieties, planting rates, planting and harvesting dates, and for obtaining effective insect, weed, and plant disease control. Thus, they

TABLE 9-4
Average Insecticide Cost Per Acre in San Joaquin Valley Citrus Orchards, 1970–1971[a,b,c]

Citrus acres	Nonusers 1970	Users 1970	Nonusers 1971	Users 1971	Nonusers 1970–1971	Users 1970–1971
0–25	32.71	7.43	53.80	14.81	43.25	21.00
	(22.27)	(3.76)	(16.19)	(14.88)	(22.29)	(21.40)
26–100	41.64	10.25	37.80	10.17	34.13	9.19
	(19.81)	(8.99)	(16.96)	(9.40)	(19.36)	(9.27)
Over 100	47.63	27.92	44.73	19.44	46.18	23.11
	(18.77)	(12.65)	(16.15)	(15.52)	(17.57)	(14.95)
Average[d]	45.64	24.58	42.97	17.99	42.35	20.53
	(19.42)	(13.89)	(16.76)	(15.14)	(18.29)	(14.97)

[a] From Hall *et al.* (1975). Data collected from Willey-Norgaard research for the Ford Foundation.
[b] Standard deviations are indicated in parentheses.
[c] "Users" are growers using consultant services; "nonusers" are growers not using consultant services.
[d] These statistics are not calculated by averaging the numbers given in the table. They are calculated using total acres and total insecticide costs for each category.

must monitor soil nutrients, weather, and crop phenology as well as pest and beneficial organism populations. They may be charged with calibrating and adjusting equipment for optimal performance and minimal pesticide waste, and generally to keep accurate records of all management practices. Since, as we have seen in earlier chapters, pest damage is often greatly influenced by weather, soil and moisture, plant variety, or various management practices, consultant groups that have specialists in all these areas are able to provide an ecosystem-based approach to pest problems.

In the future, taking such a holistic approach will be made a great deal easier through the use of computers. While crop protection specialist groups are able to assess the effects of a variety of agricultural practices on pest problems and crop yields, the integration of this information is still relatively crude. A computer programmed with the physiological needs of the crop, the biology of the pest, and the effects of weather and other environmental influences on the crop and on the major organisms living there can ascertain with far more accuracy which problems are liable to arise and what the effects of various pest control options may be on crop returns. As programs for each crop are worked out and adjusted for regional weather and problem patterns, crop protection specialists will be able to feed in specifics from their growers' acreages (e.g., pest populations, natural enemy populations, timing of irrigation, or cultural practices) and base pest control decisions on the resulting computer assessments with far more confidence than they are able to presently, using current field sampling techniques and decision-making guidelines. Already, mobile units connecting IPM specialists by telephone to university computers have been put in use in some states to provide producers with quick feedback as to whether or not a control action is needed. In California, a system of computer terminals in farm advisors' offices will soon provide the same services.

Training crop protection specialists is critical in determining the quality of work they will do. The best place for IPM specialist education is in the universities where IPM research is being done so the specialist–trainee can learn the most up-to-date methods and analysis technologies, and where he/she can observe and participate in the development of programs and make important contact with key researchers. All crop protection specialists should become familiar with all areas of crop management and keep close contact with experts in each area of specialization. The work of such specialists is essential to the perfection of IPM programs; however, the bulk of the day-to-day work of examining the fields, discussing problems with growers, and recommending pest control measures must be done by well-trained generalists who are familiar with all aspects of crop protection and know where and when to call on the specialists for advice.

The integrated pest management generalist's education should be at least to the Bachelor of Science level and embrace a well-rounded program of lectures and laboratory and field courses in plant nutrition, entomology, plant pathology, nematology, weed science, soil and water science, ecology and economics, and some exposure to computer programming and agricultural and pest monitoring techniques. The IPM specialist should be well-informed about all the crop protection techniques discussed in Chapter 8 (i.e., biological, cultural, physical, mechanical, genetic, regulatory, and chemical controls), especially as they might be used in the crops he/she manages. And he/she should be knowledgeable about the hazards associated with the use of pesticide chemicals (to people and other nontarget organisms) and be especially cognizant of the legal restrictions and safety precautions necessary for the use of any chemicals he/she might recommend. Ideally, the trainee should spend at least a year or more of postgraduate apprenticeship with an experienced IPM practitioner or group before hanging out his/her shingle and undertaking the responsibilities of decision-making as an integrated pest management specialist.

IPM specialists should be required to pass a comprehensive examination by the state or states in which they practice. The licensing examination should be rigorous to assure the licensing of only top-quality pest management advisors. Licensed pest managers should be reexamined at reasonable intervals and required to take annual "refresher" and "update" courses to acquaint themselves with the latest advancements in pest management technology in the resources they manage. Every effort should be made to make certain that those licensed to recommend use of pesticide chemicals do not also make a commission on the chemicals their advisee–clients buy. The advisement-chemical sales tie-in is, of course, a clear conflict of interest.

The position of integrated pest management specialist is a job of the future. In every area where there are pests to control, there is need for individuals knowledgeable about IPM techniques and philosophy. At times the service they provide need only be "hand-holding" to assure clients that their low pest populations are of no economic consequence. Most often, decision-making will be highly challenging, requiring a consideration of a complex of interacting biological, climatic, agricultural, and economic factors. In this job of the future, the IPM consultant will also be an educator; he/she will keep the grower or resource manager up-to-date on pest control practices in their crops or resources through discussion, reports, and literature. In urban areas, such as in the city of Berkeley (see Chapter 8), pest management specialists have already established programs to educate the general public. Thus, not only does the job of the IPM specialist require an adequate ecological and technical background, it also requires an ability to communicate effectively with all kinds of people.

Independent IPM consultant firms are growing; a recent estimate put 10% of California's irrigated acreage under IPM, while twenty years ago less than 1% of the state's acreage received such services. This is no surprise as farmers have recently become aware (through government and university-supported demonstrations, through discussion with neighbors and farm advisors, and through the performance of pest management consultants themselves) that IPM can save them time and money and maintain or even increase crop yields and produce quality. As pesticide costs rise, restrictions on the use of toxic pesticides multiply, and the complexity of agricultural and resource management technology increases, the integrated pest management specialist will certainly become a prominent figure in American agriculture and resource management.

CHAPTER 10

The Future of Integrated Pest Management

On first impression it would seem that a rapid and widescale implementation of integrated pest management is an inevitable eventuality. This is a logical assumption when we realize that in comparison to the prevailing chemical control strategy, IPM is more effective, less costly, and less hazardous to man and the environment. Logic tells us that society should be rushing to adopt this better pest management strategy, but in fact it is not. Indeed, despite the success of a variety of programs globally and the enthusiastic endorsement of IPM by a number of the world's most respected pest control researchers and practitioners, the strategy's development and implementation have moved at a snail's pace. In California, for example, where much of the pioneering effort in IPM has occurred, the strategy is only utilized on about 20% of the cotton acreage, in a fraction of the deciduous fruit and citrus orchards, in only a handful of the communities and mosquito abatement districts, and not at all on the bulk of the agricultural acreage and in other areas or resource production.

Several factors contribute to this seemingly anomalous situation. Perhaps the most important of these is the opposition of a consortium of individuals, institutions, and organizations who have a vested interest, or believe that they do, in the prevailing chemical control strategy. Here, market economics, traditional patterns of action, friendships, social and organizational associations, administrative prerogatives, political influence, and a number of other factors all combine to meld seemingly disparate individuals, groups, and organizations into a coalition opposed to change. This is typical of the human species, a creature most comfortable with what is familiar and what has traditionally been done. In pest control, the pro-status quo coalition is an extremely powerful impediment to IPM evolution, and until the log jam that it has created is broken, the expansion of IPM will be slow indeed.

Another factor that has contributed to the slow development and implementation of IPM is a widespread lack of understanding of and support for IPM research in the agencies and institutions that conduct pest control research. Researchers and administrators who really understand the IPM concept and the need for its implementation are a distinct minority. Consequently, in most federal and state experiment stations the traditional pattern of piecemeal research prevails (e.g., the orchard scene, where one person researches spider mites, another insects, a third weeds, and a fourth diseases). Most of the researchers know no other operational mode, and the same holds true for the administrators who have risen from the ranks to become policymakers. Thus there is no widescale shift in pest management research policy, and the traditional pattern prevails. In this connection, it does not help that constant pressures arise from a variety of sources for "instant" answers to nagging pest problems. Whether they are growers, mosquito-tormented urbanities, national park campers, food processors, or city folks who simply detest insects, individually and collectively, people and groups generate constant and heavy pressure for *quick* answers. As a result, enormous amounts of research energy and support funds are expended in frenetic efforts to provide "instant" solutions. Meanwhile, IPM research, which is long-term in nature and costly, is shunted aside.

Connected with all this is another factor that has slowed the IPM evolution. This is the delusion that somehow we can develop a simple all-encompassing panacea for our pest problems. We are fascinated by this prospect, and as quickly as one glittering possibility fades we seize upon another and pour energy and money into its development. As a result, since the middle 1940s an array of panaceas has come and gone, and we are left with an overall pest problem that is as big as, or bigger than, ever. Thus over the past 30 years we have witnessed the successive eras of the systemic insecticide, microbial control, pheromones, hormones, autosterilization, and now the synthetic pyrethroids. All provide bits and pieces for an overall pest management strategy, but as we expend our brainpower, energy, and funds in giving each its run at the pest complex, the IPM strategy, the proper vehicle for all of them, drags along at its snail's pace.

The pest eradication concept dear to the hearts of government pest regulatory agencies is another impediment to the expansion of IPM. In actuality, the federal government spends more money on eradication/area-control attempts against insects than it does on all entomological research. And, most ironically, the majority of these programs fail, often after the expenditure of many millions of dollars (e.g., the ill-fated fire ant and notorious barberry eradication programs). Again IPM research and implementation languish while immense amounts of money, brainpower, and energy are expended on programs that in most cases are a hopeless gamble.

Finally, IPM will not suddenly emerge as *the* pest control strategy because a wide variety of programs are not yet available, and the development of these diverse programs will take time. Furthermore, the cadre of researchers, teachers, and practitioners necessary for the development of, training in, and implementation of IPM is not at hand, and again their training and recruitment will take time.

What has just been written is perhaps pessimistic, but it is not intended as a message of despair. We are convinced that there will be an evolution to IPM if for no other reason than that society will not remain forever blind to a resource protection strategy that is better, less expensive, and less hazardous than the one currently in vogue. In fact, we anticipate that there will be an acceleration in the transition to IPM, and indeed this acceleration may already be occurring.

Several factors cause us to believe this. The first is the skyrocketing cost of pesticides, which, coupled with the snowballing pesticide usage engendered by the conventional pest control strategy, is bringing chemical control to the verge of the economic breaking point. In other words, the increasing costs of pesticides and the spiraling chemical load of the pesticide treadmill are causing conventional pest control to become so expensive that resource producers and protectors are now giving serious thought to the IPM option. On the global scene, many developing nations with limited foreign exchange resources simply cannot afford conventional chemical control, and so they are turning to IPM. This is reflected in the burgeoning IPM programs of the Food and Agricultural Organization (FAO) of the United Nations, the U.S. government's foreign aid apparatus (US-AID), and certain of the major philanthropic organizations.

Public concern over the pesticide hazard to human health and the environment and the squandering of nonrenewable natural resources (pesticides are overwhelmingly petroleum-derived) are other factors that should hasten the transition to IPM. Legislators, public health officials, consumer groups, conservationist–environmentalist organizations, public action groups, and the farm workers' representatives seem convinced that IPM can bring about a striking reduction in pesticide usage; and through litigation and legislation these groups will surely hasten the evolution. In other words, there is a widely developing societal concern over current pesticide usage and therefore the prevailing pest control strategy. As this concern widens and deepens, society will speed the transition to the integrated pest management strategy.

Bibliography

Adkisson, P. L., 1972, Use of cultural practices in insect pest management, in: *Implementing Practical Pest Management Strategies,* Proceedings of a National Extension Insect-Pest Management Workshop at Purdue University, Lafayette, Indiana, March 14–16, 1972.

Agrios, G. N., 1969, *Plant Pathology,* Academic Press, New York, 629 pp.

Allee, W. C., A. E. Emerson, O. Park, T. Park, and K. P. Schmidt, 1949, *Principles of Animal Ecology,* W. B. Saunders, Philadelphia.

Baker, K. F., and R. J. Cook, 1974, *Biological Control of Plant Pathogens,* W. H. Freeman, San Francisco.

Barnett, W. W., C. S. Davis, and G. A. Rowe, 1977, The California Integrated Pest Management Project: Summary and Evaluation. University of California Cooperative Extension, Berkeley.

Bartlett, B. R., 1958, Laboratory studies on selective aphidicides favoring natural enemies of the spotted alfalfa aphid, *J. Econ. Entomol.* **51**(3):374–378.

Batiste, W. C., A. Berlowitz, W. H. Olson, J. E. Detar, and J. L. Joos, 1973, Codling moth: Estimating time of first egg hatch in the field—A supplement to sex attractant traps in integrated control, *Environ. Entomol.* **2**(3):387–391.

Beier, M., 1973, The early naturalists and anatomists during the Renaissance and seventeenth century, in: *History of Entomology,* R. F. Smith, T. E. Mittler, and C. N. Smith (eds.), Annual Reviews, Palo Alto, California, pp. 81–94.

Berry, J. W., D. W. Osgood, and P. W. St. John, 1974, *Chemical Villains,* C. V. Mosby, St. Louis, 189 pp.

Bird, F. T., 1953, The use of a virus disease in the biological control of the European sawfly, *Neodiprion sertifer* (Geoffr.), *Can. Entomol.* **85**:437–446.

Blair, B. D., 1976, Extension pest management pilot projects—An evaluation, USDA Extension Service, ESC-579, July, 1976.

Bormann, F. H., and G. Likens, 1971, The ecosystem concept and the rational management of natural resources, *Yale Sci.* **45**(7):2–8.

Breulach, V. A., 1973, *Plant Structure and Function,* Macmillan, New York, 575 pp.

Brown, A., and R. Pal, 1971, *Insecticide Resistance in Arthropods,* World Health Organization, Geneva.

Bush, G. L., R. W. Neek, and G. B. Kitto, 1976, Screwworm eradication: Inadvertent selection for noncompetitive ecotypes during mass rearing, *Science* **193**:491–493.

California Department of Food and Agriculture, 1978, Report on the Environmental Assessment of Pesticide Regulatory Programs, Vols. I, II, and III, CDFA, Sacramento, California.

Carson, R., 1962, *Silent Spring,* Houghton Mifflin, Boston.

Clapham. W. B., Jr., 1973, *Natural Ecosystems,* Macmillan, New York, 248 pp.

Conway, G. R., 1973, Experience in insect pest modelling: A review of models, uses, and future directions, in: *Insects: Studies in Population Management,* Geier, Clark, Anderson, and Nix (eds.), Ecological Society of Australia (Memoirs 1), Canberra, pp. 103–132.

Cope, O. B., 1971, Interactions between pesticides and wildlife, *Annu. Rev. Entomol.* **16**:325–364.

Coppel, H. C., and N. F. Sloan, 1970, Avian predation, an important adjunct in the suppression of larch case bearer and introduced pine sawfly populations in Wisconsin forests, *Proc. Tall Timbers Conf.* **2**:259–272.

Council on Environmental Quality (CEQ), 1972, *Integrated Pest Management,* U.S. Government Printing Office, Washington, D. C. 41 pp.

Croft, B. A., and A. W. A. Brown, 1975, Responses of arthropod natural enemies to insecticides, *Ann. Rev. Entomol.* **20**:285.

Croft, B. A., J. L. Howes, and S. M. Welch, 1976, A computer-based extension pest management delivery system, *Environ. Entomol.* **5**(1):20–34.

DeBach, P. (ed.), 1964, *Biological Control of Insect Pests and Weeds,* Reinhold, New York.

DeBach, P., 1974, *Biological Control by Natural Enemies,* Cambridge University Press, Cambridge.

Dethier, V. G., 1976, *Man's Plague?,* Darwin Press, Princeton, New Jersey.

Doutt, R. L., and J. Nakata, 1965, Parasites for control of grape leafhopper, *Calif. Agri.* **19**:3.

Doutt, R. L., and R. F. Smith, 1971, The pesticide syndrome—Diagnosis and suggested prophylaxis, in: *Biological Control,* C. B. Huffaker (ed.), Plenum Press, New York, pp. 3–15.

Durham, W. F., and C. H. Williams, 1972, Mutagenic, teratogenic, and carcinogenic properties of pesticides, *Annu. Rev. Entomol.* **17**:123–148.

Ehrlich, P. R., and A. H. Ehrlich, 1970, *Population, Resources, Environment,* W. H. Freeman, San Francisco, 383 pp.

Ellis, H. E., M. C. Ganyard, H. M. Singletary, and R. L. Robertson, 1973, *North Carolina Tobacco Pest Management, Second Annual Report,* University of North Carolina Extension, Raleigh, 67 pp.

Embree, D. G., 1966, The role of introduced parasites in the control of the winter moth in Nova Scotia, *Can. Entomol.* **98**:1159–1168.

Embree, D. G., 1971, The biological control of the winter moth in eastern Canada by introduced parasites, in: *Biological Control,* C. B. Huffaker (ed.), Plenum Press, New York, pp. 217–226.

Epstein, S., and M. Legator, 1971, *Mutagenicity of Pesticides,* MIT Press, Cambridge, Massachusetts.

Essig, E. O., 1931, *A History of Entomology,* Hafner Publishing, New York.

Falcon, L. A., 1971, Microbial control as a tool in integrated control programs, in: *Biological Control,* C. B. Huffaker (ed.), Plenum Press, New York, pp. 346–364.

Falcon, L. A., C. Pickel, and J. White, 1976, Computerizing codling moth, *Western Fruit Grower,* January, 1976.

Fenner, F., 1965, Myxomavirus and *Oryctolagus cuniculus*: Two colonizing species, in: *Genetics of Colonizing Species*, Baber, H. C., and G. L. Stebbins (eds.), Academic Press, New York, pp. 485–499.

Flaherty, D. L., C. D. Lynn, F. L. Jensen, and M. A. Hoy, 1971, The influence of environment and cultural practices on spider mite abundance in southern San Joaquin Valley Thompson seedless vineyards, *Calif. Agri.* **25**(11):7–8.

Gallun, R. L., K. J. Starks, and W. D. Guthrie, 1975, Plant resistance to insects attacking cereals, *Annu. Rev. Entomol.* **20**:337.

Ganyard, M. C., Jr., and H. C. Ellis, 1972, A tobacco pest management pilot project in North Carolina, in: *Implementing Practical Pest Management Strategies*, Proceedings of a National Extension Insect-Pest Management Workshop, Purdue University, Lafayette, Indiana, March 14–16, 1972, pp. 132–153.

Ganyard, M. C., Jr., H. C. Ellis, and H. M. Singletary, 1972, *North Carolina Tobacco Pest Management, First Annual Report*, University of North Carolina Extension, Raleigh, 36 pp.

Gentry, C. R., W. W. Thomas, and J. M. Stanley, 1969, Integrated control as an improved means of reducing populations of tobacco pests, *J. Econ. Entomol.* **62**(6):1274–1277.

Glass, E. H. (ed.), 1975, *Integrated Pest Management: Rationale, Potential, Needs and Implementation*, Entomological Society of America Special Publication 75-2, 141 pp.

Gonzalez, D., 1970, Sampling as a basis for pest management strategies, *Proc. Tall Timbers Conf.* **2**:83–103.

Good, J. M., 1977, Progress Report on Pest Management Pilot Projects, USDA, Washington, D. C.

Graham, F., Jr., 1970, *Since Silent Spring*, Fawcett World Library, Greenwich, Connecticut, 288 pp.

Gutierrez, A. P., 1978, Applying systems analysis to integrated control, *Calif. Agri.* **32**(2):11.

Gutierrez, A. P., T. F. Leigh, Y. Wang, and R. D. Cave, 1977, An analysis of cotton production in California: *Lygus hesperus* injury—An evaluation, *Can. Entomol.* **109**:1373–1386.

Hagen, K. S., and J. M. Franz, 1973, A history of biological control, in: *History of Entomology*, R. F. Smith, T. E. Mittler, and C. N. Smith (eds.), Annual Reviews, Palo Alto, California, pp. 433–476.

Hagen, K. S., E. F. Sawall, Jr., and R. L. Tassan, 1970, The use of food sprays to increase effectiveness of entomophagous insects, *Proc. Tall Timbers Conf.* **2**:59–81.

Hagen, K. S., R. van den Bosch, and D. L. Dahlsten, 1971, The importance of naturally occurring biological control in the western United States, in: *Biological Control*, C. B. Huffaker (ed.), Plenum Press, New York, pp. 253–293.

Hall, D. C., R. B. Norgard, and P. K. True, 1975, The performance of independent pest management consultants, *Calif. Agri.* **29**(10):12–14.

Hammond, R. B., and L. P. Pedigo, 1976, Sequential sampling plans for the green cloverworm in Iowa soybeans, *J. Econ. Entomol.* **69**(2):181–185.

Harpaz, I., 1973, Early entomology in the Middle East, in: *History of Entomology*, R. F. Smith, T. E. Mittler and C. N. Smith (eds.), Annual Reviews, Palo Alto, California, pp. 21–36.

Harris, T. W., 1880, *Treatise on Some of the Insects Injurious to Vegetation*, C. L. Fling (ed.), Orange Judd, New York.

Hassall, K. A., 1969, *World Crop Protection*, Vol. 2: *Pesticides*, Iliffe Books, London.

Haynes, D. L., R. K. Brandenburg, and D. P. Fisher, 1973, Environmental monitoring network for pest management systems, *Environ. Entomol.* **2**:889.

Headley, J. C., 1975, The Economics of Pest Management, in: *Introduction to Insect Pest*

Management, R. L. Metcalf and W. H. Luckmann (eds.), John Wiley & Sons, New York, pp. 75–99.

Hensley, S. D., 1971, Management of sugarcane borer populations in Louisiana, a decade of change, *Entomophaga* **16:**133–146.

Hepp, R. E., 1976, Alternative delivery systems for farmers to obtain integrated pest management services, USDA Extension Service, Agricultural Economics Report No. 298, June 1976.

Howard, L. O., 1930, *A History of Applied Entomology,* Smithsonian Institution, Washington, D. C.

Howell, F. J., 1972, An improved sex attractant trap for codling moths, *J. Econ. Entomol.* **65**(2)**:**609–610.

Hoyt, S. C., 1969, Integrated chemical control of insects and biological control of mites on apple in Washington, *J. Econ. Entomol.* **62:**74–86.

Hoyt, S. C., and E. C. Burts, 1974, Integrated control of fruit pests, *Annu. Rev. Entomol.* **19:**231–252.

Hudon, M., 1968, Minimum number of insecticide applications for the control of the European corn borer on sweet corn in Quebec, *J. Econ. Entomol.* **61:**75–78.

Huffaker, C. B. (ed.), 1971, *Biological Control,* Plenum Press, New York.

Huffaker, C. B., and B. A. Croft, 1976, Integrated pest management in the U.S.: Progress and promise, *Environ. Health Persp.* **14:**167–183.

Huffaker, C. B., and C. E. Kennett, 1956, Experimental studies on predation: Predation and cyclamen mite populations on strawberries in California, *Hilgardia* **26:**191–222.

Huffaker, C. B., and P. S. Messenger, 1976, *Theory and Practice of Biological Control,* Academic Press, New York.

Hunt, E. G., and A. I. Bischoff, 1960, Inimical effects on wildlife of periodic DDD applications to Clear Lake, *Calif. Fish and Game* **46:**91–106.

ICAITI, 1976, An environmental and economic study of the consequences of pesticide use in Central American cotton production. Final Report (Phase 1), Guatemala, 222 pp.

Jensen, F. L., C. D. Lynn, E. M. Stafford, and H. Kido, 1965, Insecticides for control of grape leafhopper, *Calif. Agri.* **19**(4)**:**10–11.

Johnson, H. E., and C. Pecor, 1969, Coho salmon mortality and DDT in Lake Michigan, *Trans. 34th North American Wildl. Nat. Res. Conf.,* Washington, D. C., March 3–5.

Jones, D. P., 1973, Agricultural entomology, in: *History of Entomology,* R. F. Smith, T. E. Mittler, and C. N. Smith (eds.), Annual Reviews, Palo Alto, California, pp. 307–332.

Klingman, G. E., 1961, *Weed Control as a Science,* John Wiley & Sons, New York.

Kogan, M., 1975, Plant resistance in pest management, in: *Introduction to Insect Pest Management,* R. L. Metcalf and W. Luckmann (eds.), John Wiley & Sons, New York, pp. 103–146.

Konishi, M., and Y. Ito, 1973, Early entomology in East Asia, in: *History of Entomology,* R. F. Smith, T. E. Mittler, and C. N. Smith (eds.), Annual Reviews, Palo Alto, California, pp. 1–20.

Krebs, C., 1972, *Ecology,* Harper & Row, New York, 694 pp.

Lehane, B., 1969, *The Compleat Flea,* Viking Press, New York.

Leius, K., 1967, Influence of wild flowers on parasitism of tent caterpillar and codling moth, *Can. Entomol.* **99:**444–446.

Lord, F. T., 1947, The influence of spray programs on the fauna of apple orchards in Nova Scotia. II. Oystershell scale, *Lepidosaphes ulmi* (L.), *Can. Entomol.* **79:**196–209.

Luck, R. F., R. van den Bosch, and R. Garcia, 1977, Chemical insect control, a troubled pest management strategy, *BioScience* **27:**606–611.

Luckmann, W., and R. L. Metcalf, 1975, The pest management concept, in: *Introduction to Insect Pest Management,* R. L. Metcalf and W. Luckmann (eds.), John Wiley & Sons, New York, pp. 3–36.

MacPhee, A. W., and C. R. MacLellan, 1971, Ecology of apple orchard fauna and development of integrated pest control in Nova Scotia, *Proc. Tall Timbers Conf.* **3**:197–208.

MacPhee, A. W., and C. R. MacLellan, 1971, Cases of naturally occurring biological control in Canada, in: *Biological Control*, C. B. Huffaker (ed.), Plenum Press, New York, pp. 312–328.

Maddox, J. V., 1975, Use of Diseases in Pest Management, in: *Introduction to Insect Pest Management*, R. L. Metcalf and W. Luckmann (eds.), John Wiley & Sons, New York, pp. 189–234.

Mellanby, K., 1967, *Pesticides* and *Pollution*, Collins, London, 219 pp.

Metcalf, R. L., 1975, Insecticides in pest management, in: *Introduction to Insect Pest Management*, R. L. Metcalf and W. Luckmann (eds.), John Wiley & Sons, New York, pp. 235–274.

Metcalf, R. L., and W. Luckmann (eds.), 1975, *Introduction to Insect Pest Management*, John Wiley & Sons, New York, 587 pp.

Metcalf, R. L., and R. A. Metcalf, 1975, Attractants, repellents and genetic control in pest management, in: *Introduction to Insect Pest Management*, R. L. Metcalf and W. Luckmann (eds.), John Wiley & Sons, New York, pp. 275–307.

Michelbacher, A E., and R. F. Smith, 1943, Some natural factors limiting abundance of the alfalfa butterfly, *Hilgardia* **15**:369–397.

Miller, T. D., and M. N. Schroth, 1972, Monitoring the epiphytic populations of *Erwinia amylovora* on pear with a selective medium, *Phytopathology* **62**:1175–1182.

Morge, G., 1973, Entomology in the western world in Antiquity and Medieval Times, in: *History of Entomology*, R. F. Smith, T. E. Mittler, and C. N. Smith (eds.), Annual Reviews, Palo Alto, California, pp. 37–80.

Morris, R. F., 1960, Sampling insect populations, *Annu. Rev. Entomol.* **5**:243–264.

Nash, R. G., and E. A. Woolson, 1967, Persistence of chlorinated hydrocarbons in soils, *Science* **157**:924–927.

National Academy of Sciences, 1969, *Insect-Pest Management and Control*, Principles of Plant and Animal Pest Control, Vol. 3, NAS, Washington, D. C.

National Academy of Sciences, 1975, *Cotton Pest Control*, NAS, Washington, D. C., 139 pp.

Neilson, W. T. A., N. A. Patterson, and A. D. Pickett, 1968, Field and laboratory studies for control of the apple maggot in Nova Scotia, *J. Econ. Entomol.* **61**:802–805.

Odum, E. P., 1969, The strategy of ecosystem development, *Science* **164**:262–270.

Odum, E. P., 1971, *Fundamentals of Ecology*, 3rd ed., W. B. Saunders, Philadelphia, 574 pp.

Olkowski, W., 1973, A model ecosystem management program, *Proc. Tall Timbers Conf.* **5**:103–117.

Olkowski, H., and W. Olkowski, 1976, Entomophobia in the urban ecosystem, some observations and suggestions, *Bull. Entomol. Soc. Am.* **22**(3):313–317.

Olkowski, W., H. Olkowski, R. van den Bosch, and R. Hom, 1976, Ecosystem management: A framework for urban pest control, *BioScience* **26**(6):384–389.

Ordish, G., 1976, *The Constant Pest*, Charles Scribner's Sons, New York.

Patterson, N. A., 1966, The influence of spray programs on the fauna of apple orchards in Nova Scotia. XVI. The long-term effect of mild pesticides on pests and their predators, *J. Econ. Entomol.* **59**:1430–1435.

Pinnock, D. E., R. J. Brand, J. E. Milstead, and N. F. Coe, 1974, Suppression of populations of *Aphis gossypii* and *A. spiraecola* by soap sprays, *J. Econ. Entomol.* **67**(6):783–784.

Prokopy, R. J., 1968, Sticky spheres for estimating apple maggot adult abundance, *J. Econ. Entomol.* **61**(4):1082–1085.

Prokopy, R. J., 1975, Apple maggot control by sticky red spheres, *J. Econ. Entomol.* **68**(2):197–198.

Rabb, R. L., 1969, Environmental manipulations as influencing populations of tobacco hornworms, *Proc. Tall Timbers Conf.* **1**:175–191.

Rabb, R. L., 1971, Naturally occurring biological control in the eastern U.S. with particular reference to tobacco insects, in: *Biological Control,* C. B. Huffaker (ed.), Plenum Press, New York, pp. 294–311.

Rabb, R. L., and F. E. Guthrie (eds.), 1970, *Concepts of Pest Management,* North Carolina State University Press, Raleigh, North Carolina, 242 pp.

Rabb, R. L., H. H. Neunzig, and H. V. Marshall, Jr., 1964, Effect of certain cultural practices on the abundance of tobacco hornworms and corn earworms in tobacco after harvest, *J. Econ. Entomol.* **57**(5):791–792.

Rabb, R. L., F. A. Todd, and H. C. Ellis, 1974, Tobacco Pest Management, in: AAAS Symposium: Pest Management—An Interdisciplinary Approach to Crop Protection, San Francisco, February 28, 1974.

Ratcliffe, D. A., 1967, Decrease in eggshell weight in certain birds of prey, *Nature* **215**:208–210.

Reynolds, H. T., P. L. Adkisson, and R. F. Smith, 1975, Cotton insect pest management, in: *Introduction to Insect Pest Management,* R. L. Metcalf and W. Luckmann (eds.), John Wiley & Sons, New York.

Risebrough, R. W., R. J. Huggett, J. J. Griffin, and E. D. Goldberg, 1968, Pesticides: Transatlantic movements in the northeast trades, *Science* **159**:1233–1236.

Robbins, W. W., A. S. Crafts, and R. N. Raynor, 1942, *Weed Control,* McGraw-Hill, New York.

Robbins, W. W., M. K. Bellue, and W. S. Ball, 1970, *Weeds of California,* California Department of Agriculture, Sacramento, 547 pp.

Ruesink, W. G., 1975, Analysis and modeling in pest management, in: *Introduction to Insect Pest Management,* R. L. Metcalf and W. Luckmann (eds.), John Wiley & Sons, New York, pp. 353–378.

Ruesink, W. G., and M. Kogan, 1975, The quantitative basis of pest management: Sampling and measuring, in: *Introduction to Insect Pest Management,* R. L. Metcalf and W. Luckmann (eds.), John Wiley & Sons, New York, pp. 309–352.

Salisbury, F. B., and C. Ross, 1969, *Plant Physiology,* Wadsworth, New York, 747 pp.

Sanderson, E. D., 1915, *Insect Pests of Farm, Garden and Orchard,* John Wiley & Sons, New York.

Smith, H. S., 1941, Racial segregation in insect populations and its significance in applied entomology, *J. Econ. Entomol.* **34**:1–13.

Smith, R. F., 1971, Economic aspects of pest control, *Proc. Tall Timbers Conf.* **3**:53–83.

Smith, R. F., and K. S. Hagen, 1959, Impact of commercial insecticide treatments, *Hilgardia* **29**(2):131–154.

Smith, R. F., and R. van den Bosch, 1967, Integrated control, in: *Pest Control,* R. L. Doutt (ed.), Academic Press, New York, pp. 295–340.

Smith, R. F., T. E. Mittler, and C. N. Smith (eds.), 1973, *History of Entomology,* Annual Reviews, Palo Alto, California.

Southwood, T. R. E., 1966, *Ecological Methods,* Methuen, London, 391 pp.

Southwood, T. R. E., and G. A. Norton, 1973, Economic aspects of pest management strategies and decisions, in: *Insects: Studies in Population Management,* Geier, Clark, Anderson, and Nix (eds.), Ecological Society of Australia (Memoirs 1), Canberra, pp. 168–184.

Stern, V., 1973, Economic thresholds, *Annu. Rev. Entomol.* **18**:259–280.

Stern, V. M., and R. van den Bosch, 1959, The integration of chemical and biological control of the spotted alfalfa aphid. II. Field experiments on the effects of insecticides, *Hilgardia* **29**(2):103–130.

Stern, V. M., R. F. Smith, R. van den Bosch, and K. S. Hagen, 1959, The integration of chemical and biological control of the spotted alfalfa aphid. I. The integrated control concept, *Hilgardia* **29**(Z):81–101.

Strickland, A. H., 1961, Sampling crop pests and their hosts, *Annu. Rev. Entomol.* **6**:201–220.

Summers, C. G., 1976, Population fluctuations of selected arthropods in alfalfa: Influence of two harvesting practices, *Environ. Entomol.* **5**(1):103–110.

Teetes, G. L., and N. M. Randolph, 1971, Effects of pesticides and dates of planting sunflowers on the sunflower moth, *J. Econ. Entomol.* **64**:124–126.

Telford, A. D., 1977, A case of integrated pest management against mosquitoes, in: *New Frontiers in Pest Management,* California State Assembly, Sacramento, pp. 61–63.

U.S. Department of Health, Education, and Welfare, 1969, *Report of the Secretary's Commission on Pesticides and Their Relationship to Environmental Health,* E. M. Mrak, (Chairman), DHEW, Washington, D. C., 46 pp.

van den Bosch, R., and P. S. Messenger, 1973, *Biological Control,* Intext, New York, 180 pp.

van den Bosch, R., E. I. Schlinger, E. J. Dietrick, K. S. Hagen, and J. K. Halloway, 1959, The colonization and establishment of imported parasites of the spotted alfalfa aphid in California, *J. Econ. Entomol.* **52**(1):136–141.

van den Bosch, R., T. F. Leigh, L. A. Falcon, V. W. Stern, D. Gonzales, and K. S. Hagen, 1971, The developing program of integrated control of cotton pests in California, in: *Biological Control,* C. B. Huffaker (ed.), Plenum Press, New York, pp. 377–394.

van Emden, H. F., 1966, Plant–insect relationships and pest control, *World Rev. Pest Control* **5**:115–123.

Varley, G. C., G. R. Gradwell, and M. P. Hassell, 1974, *Insect Population Ecology,* University of California Press, Berkeley, 212 pp.

von Rumker, R., and G. L. Kelso, 1975, A study of the efficiency of the use of pesticides in agriculture. A Report for the EPA, EPA-540/9-75-025.

von Rumker, R., R. M. Matter, D. P. Clement, and F. K. Erickson, 1972, The use of pesticides in suburban homes and gardens and their impact on the aquatic environment. Pesticide Study Series 2, EPA Office of Water Programs, Applied Technology Division, Washington, D. C.

von Rumker, R., E. W. Lawless, and A. F. Meiners, 1974, Production, distribution, use, and environmental impact potential of selected pesticides, EPA Report 540/1-74-001, EPA, Washington, D. C., 439 pp.

Waddill, W. H., B. M. Shepard, S. G. Turnispeed, and G. R. Carner, 1974, Sequential sampling plans for *Nabis* spp. and *Georcoris* spp. on soybeans, *Environ. Entomol.* **3**:415–419.

Ware, G., 1975, *Pesticides, An Auto-Tutorial Approach,* W. H. Freeman, San Francisco, 191 pp.

Warren, G. L., 1970, Introduction of the masked shrew to improve control of forest insects in Newfoundland, *Proc. Tall Timbers Conf.* **2**:185–202.

Waters, W. E., 1955, Sequential sampling in forest insect surveys, *For. Sci.* **1**:68–79.

Watson, T. F., L. Moore, and G. Ware, 1975, *Practical Insect Pest Management,* W. H. Freeman, San Francisco, 196 pp.

Watt, K. E. F., 1974, *Ecology,* Biocore Unit XXI, McGraw-Hill, New York, 23 pp.

Weires, R. W., and H. C. Chiang, 1973, Integrated control prospects of major cabbage insect pests in Minnesota, University of Minnesota Agricultural Experiment Station Technical Bulletin **291**, 42 pp.

Whittaker, R. H., 1975, *Communities and Ecosystems,* 2nd ed., Macmillan, New York, 385 pp.

Wilson, C. L., and W. B. Loomis, 1962, *Botany*, Holt, Rinehart and Winston, New York.

Wood, B. J., 1973, Integrated control: Critical assessment of case histories in developing economies, in: *Insects: Studies in Population Management*, Geier, Clark, Anderson, and Nix (eds.), Ecological Society of Australia (Memoirs 1), Canberra, pp. 196–220.

Woods, A., 1974, *Pest Control: A Survey*, John Wiley & Sons, New York, 407 pp.

Woodwell, G. M., 1967, Toxic substances and ecological cycles, *Sci. Am.* **216**(3):24–31.

Woodwell, G. M., C. F. Wurster, and P. A. Isaacson, 1967, DDT residues in an estuary: A case of biological concentration of a persistent insecticide, *Science* **156**:821–824.

Wurster, C. F., 1972, Effects of insecticides, in: *The Environmental Future*, N. Polunin (ed.), Macmillan, London, pp. 293–310.

Wurster, C. F., Jr., and B. Wingate, 1968, DDT residues and declining reproduction in the bermuda petrel, *Science* **159**:979–981.

Index